EXACT CONSTRAINT:
MACHINE DESIGN USING KINEMATIC PRINCIPLES

Douglass L. Blanding

ASME Press   **New York**   **1999**

Copyright © 1999 by   The American Society of Mechanical Engineers
                        Three Park Avenue, New York, NY 10016

Library of Congress Cataloging-in-Publication Data

Blanding, Douglass L., 1950 –
Exact Constraint: Machine Design Using Kinematic Processing / Douglass L.
Blanding
p. cm.
ISBN 978-0-7918-0085-0
1. Machine design.   2. Machinery, Kinematics of.   I. Title.

TJ230.B593 1999   99-
043043
621.8'16 21 — dc21
CIP

Contents

Preface

Exact Constraint: Machine Design Using Kinematic Principles explores the venerable but somewhat obscure principles of kinematic design (widely practiced in the design of precision instruments for well over 100 years) and knits them into a unique and powerful set of rules and techniques that can be used to facilitate the design of any machine. These rules and techniques apply to the design of machines of every type and size, not just to the design of precision instruments. A central technique is Constraint Pattern Analysis, which enables a designer to visualize the constraints and degrees of freedom of a mechanical connection as patterns of lines in space. When the spatial relationships among these patterns of lines are explored, they are found to obey some fairly simple "rules"—rules that are not new, but which are nevertheless not well known. These patterns of lines are found in all types of machinery, from precision instruments to office products to large vehicles. By learning to recognize these line patterns in machines, the mechanical design engineer is then able to understand the way a machine will work (or will not work) in an entirely new domain. This can often lead an engineer to a unique and unobvious solution to his or her design problem. The use of these principles, collectively called Exact Constraint Design principles, provides the designer with a better understanding of a machine's behavior. This understanding allows the designer to easily create new designs which are both low in cost and high in performance.

Exact Constraint Design principles unite areas of machine design that seem to be categorically different. For example, patterns of lines representing constraints and freedoms can be found in the connections between the parts of a mechanism as well as within a single structure. Thus, such disparate fields as structural analysis and mechanism design turn out to be categorically similar. The same analysis techniques apply equally well to both areas.

Exact Constraint Design principles apply whenever mechanical connections exist between parts of a machine. Sometimes, we must look beyond the boundaries of the machine itself to identify these connections. For example, the tires of an automobile are "connected to" the road. Engineers who design the suspension of an automobile must take account of this connection in order to determine geometry characteristics such as the location of a car's roll axis. Similarly, in a factory machine that conveys long, thin webs of material, such as paper, we must treat the web itself as a component part of the machine and consider the nature of the mechanical connection between the web and the rollers on which it is being conveyed. The final chapter of this book is devoted to the application of Exact Constraint Design principles to web conveyance apparatus.

This book is for the aspiring machine designer. It is for engineers, designers, scientists, technicians, and mechanics who are involved in the design of machines: automobiles, farm machinery, airplanes, space stations, telescopes, office equipment, appliances, and so on. It is for anyone who is fascinated by machinery and who wants to better understand how machines work. The subject is developed from the basic fundamentals of kinematics. The abstract theory is supplemented with many hardware examples. The reader will acquire a greatly enhanced understanding of how machines work, an improved ability to create innovative machines with high performance and low cost, and the ability to analyze and understand complex existing machinery.

It is the author's purpose in writing this book to present a collection of rules and techniques pertaining to kinematic aspects of machine design and to assemble them into a tidy, practical body of knowledge that will enable a practicing design engineer to do his or her job better. Some of these rules are well known. Some are not. Some of the techniques are new. The technique of Constraint Pattern Analysis is new. Certainly, the collection of all of these rules and techniques, assembled in this manner, is new. Collectively called the Exact Constraint Design Principles, this is a practical methodology that takes the mystery out of kinematics and enables kinematic design principles to be applied over a wide range of mechanical design problems, not just to the design of precision instruments.

This book presents the principles of Exact Constraint Design by first introducing the fundamental concepts and then gradually building on those concepts, using familiar hardware examples. It is hoped that this approach will make the abstract concepts seem more tangible. This book is *not* intended to be a catalog of kinematic connections. Many clever examples of kinematic connections are presented. The author neither claims credit for devising these examples nor accuracy in ascribing them to their "original" inventor. These examples are presented to illustrate various principles of Exact Constraint Design.

Background

Although he did not claim to be the first to discover the principles of kinematics, James Clerk Maxwell, a Scottish physicist, gave a succinct summary of what is meant by a kinematic design solution. In his paper "General Considerations Concerning Scientific Apparatus,"[1] he outlined that the number of mechanical constraints applied to a critical component of a scientific instrument must equal six minus the number of degrees of freedom that component is intended to have. Furthermore, he pointed out that these constraints must be properly placed in order to avoid overconstraint. The roots and evolution of kinematic design theory are traced by Chris Evans in his book *Precision Engineering: An Evolutionary View.*[2]

The use of kinematic design principles has come to be regarded as an essential discipline in the design of precision instruments. T. N. Whitehead, in his book *The Design and Use of Instruments and Accurate Mechanism,*[3] presented precision kinematic design as an essential requirement in the design of instruments and accurate mechanisms. Among the benefits to be attained by following these principles are extreme precision, predictable performance, and infinitesimal distortions of component parts.

In the 1960s, Dr. John McLeod of the Eastman Kodak Company used the principles of kinematic design in his work to design rigid structures and flexural mounts for precision optical components. Flexures are thin (usually metal) sheets or wires that sustain moderate loads in tension or compression with only negligible distortion, but bend very easily under the application of transverse or bending loads. Dr. McLeod showed how wire flexures can be connected in various patterns to achieve different constraint conditions, and therefore different degrees of freedom. He also observed that the sheets and bars of structures, despite being quite thick, are similar in shape to flexures and will therefore have a similar response to applied loads. This observation opened the way to extend the use of kinematic design principles to the design of structures. Dr. McLeod coined the term "exact constraint" to describe the condition when a component of interest has been neither underconstrained nor overconstrained, but constrained by a "kinematically correct" pattern of constraints, providing exactly the degrees of freedom that are desired.

The use of these design principles was further extended by John E. Morse, also of the Eastman Kodak Company. An electrical engineer by training, Mr. Morse became involved in Exact Constraint Design by an unlikely series of events. He was aware of Dr. McLeod's work and his explanations of overconstraint, underconstraint, and exact constraint in structures and flexure mechanisms for mounting optical components. At the time, Mr. Morse was trying to solve a seemingly unrelated web conveyance problem. He then made the profound observation that the web being conveyed by his machine was in fact a two-dimensionally rigid part of the machine and that it was being *overconstrained* in its

[1] Maxwell, J. C. *The Scientific Papers of James Clerk Maxwell* (vol. 2). Cambridge University Press, London, 1890.

[2] Evans, C. J. *Precision Engineering: An Evolutionary View.* Cranfield Press, Bedford, UK, 1989.

[3] Whitehead, T. N. *The Design and Use of Instruments and Accurate Mechanism.* Macmillan, New York, 1934.

connection to the machine. By making this observation, Mr. Morse was able to extend the principles of kinematic design to analyze and solve his web conveyance problem. Mr. Morse then continued to develop and refine this technique, which he called "exact constraint web handling." During the early 1980s, Mr. Morse energetically promoted the use of Exact Constraint Design throughout the Eastman Kodak Company, delivering many tutorial lectures and publishing two comprehensive sets of notes: *Exact Constraint Machine Design* and *Exact Constraint Web Handling*.

I had the good fortune to work with Mr. Morse from 1984 until his retirement in 1986. During that time, I became a great believer in the power of using kinematic design principles to solve design problems in many areas of machine design. I have continued to work in this field and have published (internally at Kodak) numerous short papers documenting notable Exact Constraint solutions to design problems that are of general interest to engineers throughout the Eastman Kodak Company. Recently, in response to numerous requests, I undertook the task of consolidating the various written materials on this subject. I started with the intention to organize the existing material into a format that would consolidate the presentation of the fundamentals, using examples to show the relation between constraints and freedoms in a way that is both easy to understand and easy to apply by anyone with an interest in the subject.

I struggled to find a way to explain how to arrive at certain solutions. There was no *method* that would always reveal the degrees of freedom of a body resulting from the application of an arbitrary pattern of constraints, especially when those constraints were not aligned along orthogonal directions. This is the sort of thing that a good designer might develop an intuition for, with years of experience, but then still be unable to convey clearly to someone who is inexperienced. I also found that even experienced designers had trouble understanding and analyzing mechanical connections having constraint patterns they had not seen before. An example of this came to my attention during a tutorial lecture I attended on the subject of precision kinematic design. In this tutorial, an apparatus was described that contained a mechanical connection between two bodies consisting of a ball-and-socket joint and three micrometer screw connections along skew lines. The lecturer, who was experienced at kinematic design, was at a loss to describe the exact degrees of freedom controlled by each of the three screws. It was clear to me that a piece of the puzzle was missing.

I finally found that missing piece when I used six pure rotations ($\mathbf{R}$s) to represent a body's six degrees of freedom instead of the customary three translations plus three rotations. This enabled the discovery of a simple relationship between the pattern of lines representing a body's freedoms and the pattern of lines representing the mechanical constraints applied. This relation is the central theme of Constraint Pattern Analysis. This is the piece that had been missing. Ironically, it turned out not to be new. It was just very obscure. Nevertheless, now there was a method that could neatly explain all the various "clever solutions" that only seemed obvious in retrospect. This was the "glue" that gave cohesiveness to all of the different areas where kinematic design had been seen to apply. This book presents these principles and techniques organized into a fairly concise format, which should prove to be a unique and valuable resource to anyone involved in machine design.

Whereas most engineering books tend to be *quantitative* in nature, the principles and techniques set forth in this book are *qualitative*. The material in this book is not intended to compete with or supplant any engineering analysis. To the contrary, Exact Constraint Design principles are intended to help a designer come up with the proper *topology* of a machine, whether it is a mechanism or a structure or both. They are intended to guide a designer to the correct location and placement of mechanical constraints to achieve the degrees of freedom desired. Thus, this book teaches a *design methodology*. The material in this book should be learned in conjunction with sound engineering fundamentals.

Acknowledgments

I thank John E. Morse for igniting my interest in this subject more than a decade ago. It is certainly true that this subject received an enormous boost as a result of Mr. Morse's uniquely clever insights regarding Exact Constraint Design. I am fortunate to have had an opportunity to work under his tutelage.

CHAPTER *1*

Two-Dimensional Connections Between Objects

We are about to explore the nature of mechanical connections between objects. Those objects have, in their free state, six independent degrees of freedom of motion or position: three translational and three rotational. When we design the mechanical connections between parts of a machine, we must account for all six of these degrees of freedom. If we do a good job of this, it will help us to efficiently design machines that are high performance and low cost.

Our exploration begins with the most fundamental principles, starting with two-dimensional connections, where we need to consider only three degrees of freedom. We will also use some simple models. Just as "a picture is worth a thousand words," it is also true that "a model is worth a thousand pictures." The reader is encouraged to construct the various simple models described in the text.

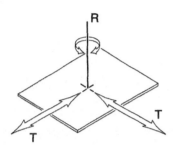

FIGURE 1.1.1

1.1 ■ DEGREES OF FREEDOM

In three dimensions (3D), an object has exactly six independent degrees of freedom of motion. However, in two dimensions (2D), an object has only three degrees of freedom: two translational and one rotational. As a 2D example, consider a sheet of cardboard (such as the back of a pad of paper) resting on a flat surface (such as a table top). Assuming the cardboard will remain in contact with the table surface, it has just three independent degrees of freedom of motion with respect to the table:

1. Right to left translation
2. Front to back translation
3. Rotation about an axis perpendicular to the table surface

The symbol T is used to designate a translational degree of freedom.

The symbol R is used to designate a rotational degree of freedom.

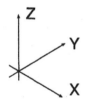

FIGURE 1.2.1

1.2 ■ COORDINATE AXES

Often, it is convenient to refer to these degrees of freedom with respect to a set of coordinate axes which is used to establish a frame of reference. In 3D, we will use the conventional rectangular coordinate system where X, Y, and Z are orthogonal. In our 2D example, we name the "right-to-left" translational degree of freedom (parallel to the X-axis) the "X degree of freedom." The "front-to-back" translational degree of freedom (parallel to the Y-axis) is called the "Y degree of freedom." The rotational degree of freedom is called "θ_z" because it is rotational about a line parallel to the Z-axis.

1.3 ■ CONSTRAINTS

When a mechanical connection is made between an object and a reference object in such a way that the number of degrees of freedom of the object (relative to the reference object) is *reduced*, we say that the object has been *constrained*.

There is a one-to-one correspondence between the constraints applied to an object and the degrees of freedom removed. For example, a 2D body, having three degrees of freedom in its free state, would end up with only two degrees of freedom remaining as a result of the application of a single constraint. Likewise, a 2D body with two constraints applied would be left with only one degree of freedom. Three constraints would result in zero degrees of freedom.

Now let us return to our 2D model, the piece of cardboard on the table. Suppose we wish to remove the X degree of freedom. We can do this by applying a single X constraint, as shown in Fig. 1.3.1.

Here, we have modeled a constraint by attaching a cardboard *link* using two thumbtacks. One end of the link is attached to our 2D object and the other end is attached to our work surface. We have thus applied a constraint to the object. The axis of the link (the line joining the two tacks) defines what is called the *constraint line*. The constraint has this effect:

> Points on the object along the constraint line can move only at right angles to the constraint line, not along it.

Because the constraint line of the link is parallel to the X-axis, this constraint removes the object's X degree of freedom. It is called an X constraint. Now, if we check the model, we see that indeed the object is no longer free to move in X, but that it still has freedom to move in θ_z and Y (for small motions).

Notice that the object does not have a large or infinite range of motion in Y. It is free in Y "at the instant" it is in the position shown. Analogously, imagine that we are asked to find the speed of an accelerating car just as the car drives past us. We find the speed to be 25 mph just as the car passes. Earlier, the car's speed was less than 25. After it passed, its speed was greater than 25. Twenty-five mph was its instantaneous speed at the *moment of interest*. Likewise, we say that our model of Fig. 1.3.1 has θ_z and Y degrees of freedom at the moment it is in the position shown.

Now let us examine some other constraint devices that perform exactly the same function as the link.

Figure 1.3.2 shows a bar connection to the object. With this device, the bar axis defines the constraint line. The bar can be modeled with a ⅛-inch diameter wooden dowel connected at the ends with a dab of hot-melt glue. One end is connected to the object and the other end is connected to the table or work surface. Hot-melt glue has the characteristic of being "rubbery" enough to permit small angular displace-

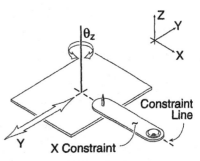

FIGURE 1.3.1

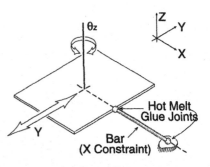

FIGURE 1.3.2

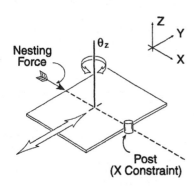

FIGURE 1.3.3

ment at the joints while still providing a very stiff connection to forces along the constraint line.

Another constraint device is the simple contact point shown in Fig. 1.3.3. This shows a short post extending from the table surface against which the object is held in contact by a *nesting force*. The nesting force must be of a magnitude that will ensure that contact is maintained between the object and the post throughout any Y or θ_z excursion of the object, and regardless of any other incidental loads that might be applied. For this device, the constraint line is defined as the line perpendicular to the contact surfaces at the point of contact.

Regardless of which of these constraint devices we select—the link, the bar, or the contact point—the remaining degrees of freedom of the body are the same. The body will have a rotational degree of freedom (**R**) intersecting the constraint line and perpendicular to the plane of the 2D body; and a translational degree of freedom (**T**) perpendicular to the constraint line and lying in the plane of the 2D body. These two degrees of freedom are independent of each other. That is, the body can rotate (through a small excursion) about any given point along the constraint line, or this same point can be made to move translationally (through a small excursion). Either motion can take place with or without the other motion. Because these two motions can be made independently, the body is said to have two degrees of freedom.

After examination of the behavior of the 2D object constrained with each of the three devices described—the link, the bar, and the contact point—we can conclude that they are all functionally equivalent. Each provides a single constraint to the object. For small motions, the object's remaining degrees of freedom are the same regardless of which constraint device is used.

In other words, when we consider an object's degrees of freedom, it is not important *which* constraint device is used, only *whether* a constraint device is used and *where*. We represent a constraint using the symbol shown in Fig. 1.3.4. This can be thought of as a bar with articulating joints at the ends to connect between objects. The bold uppercase letter **C** is used to refer to a constraint.

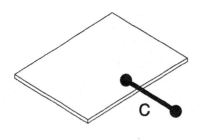

FIGURE 1.3.4

1.4 ■ FUNCTIONAL EQUIVALENCE OF CONSTRAINTS ALONG A GIVEN LINE

Suppose the link (constraint) shown in Fig. 1.3.1 had been applied along the same constraint line but to the *opposite* side of our 2D object, as in Fig. 1.4.1. What effect would this have?

Or suppose the link were longer and attached *across* our object to the far side, as in Fig. 1.4.2, but still along the same constraint line.

Does it matter how long the link is or where on the body it attaches, or from which direction it connects along a given constraint line? For small motions, the answer is no. The two remaining degrees of freedom of the body are the same for any constraint applied along a given constraint line. This observation allows us to postulate:

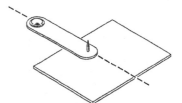

FIGURE 1.4.1

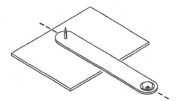

FIGURE 1.4.2

> Any constraint along a given constraint line is functionally equivalent to any other constraint along the same constraint line (for small motions).

1.5 ■ OVERCONSTRAINT

Suppose we were to install two constraints simultaneously along the same constraint line, as shown in Fig. 1.5.1, A or B. This produces a condition that is called *overconstraint*. In either case (A or B), two constraints are competing to control the same degree of freedom (X in this case).

Overconstraint results in several practical difficulties. If the dimensions of the object or the constraints are not *just right*, the parts will not fit properly. They will be either too loose or they will bind and interfere. So, we are faced with a choice between parts that have loose, sloppy, imprecise location, or parts that do not go together because they are too tight and do not fit. If we are willing to spend more money, we can opt for a third choice: get parts that fit together perfectly (or nearly so). To do this, we can either tighten the tolerances on the fit dimensions, or we can use special assembly techniques such as drilling and pinning in place. But, even if we get parts that fit together perfectly, there is another penalty that goes with overconstraint: the buildup and transfer of stresses.

Consider the configuration shown in Fig. 1.5.3 where body A is overconstrained to body B in X. Assume that the connections have been made either

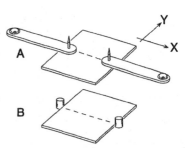

FIGURE 1.5.1

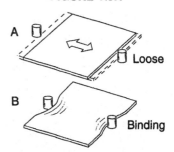

FIGURE 1.5.2

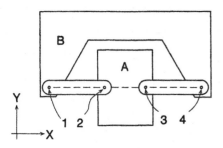

FIGURE 1.5.3

by drilling and pinning during assembly or through the use of parts whose dimensions have been very accurately controlled.

Now, when the dimension from 1 to 4 changes slightly as a result of a temperature change or some other externally induced distortion of body B, internal forces are generated along the constraint line, causing a buildup of stress in both bodies and in the links that connect them. If this stress becomes too high, approaching the material's yield strength, it can actually weaken the assembly and result in premature failure. In fact, the strength of the assembly can actually be *less* than it would have been if only *one* link had been used.

To summarize, overconstraint generally results in reduced performance (binding, stress, slop, imprecision), higher cost (tight tolerances, special assembly techniques), or both. For these reasons, it is a condition that we should always try to recognize and generally try to avoid.

1.6 ■ INSTANT CENTER OF ROTATION (VIRTUAL PIVOT)

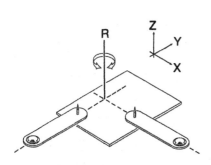

FIGURE 1.6.1

Suppose now that we wish to remove a second degree of freedom from our 2D object. This can be done by applying a second link, as in Fig. 1.6.1.

Symbolically, we can represent this situation as shown in Fig. 1.6.2. With the application of these two constraints, both the X and Y degrees of freedom of the object are removed. The object now has only one remaining degree of freedom. This is a rotational degree of freedom about a line whose location is uniquely determined to be at the intersection of the X and Y constraint lines and perpendicular to the X–Y plane. This line can also be referred to as a *virtual pivot* or an *instant center of rotation*, the latter name suggests that its instantaneous location shifts as the object moves. As the object moves, the constraint lines also move, and therefore, their point of intersection shifts. In many applications, where an object undergoes only a small motion, the magnitude of this shift is inconsequential.

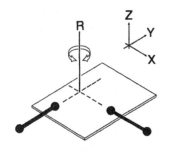

FIGURE 1.6.2

The explanation of how two intersecting constraints define a pivot is as follows: Recall (from section 1.3) that the effect of a constraint is to permit points along the constraint line to move only in a direction perpendicular to the constraint line, not along it. This requires that the body must be *rotating* about a point somewhere along that constraint line. When two constraints are applied to a 2D body, the body must rotate about a point that is on *both* constraint lines. The intersection of the two lines is the only point that lies on both constraint lines, therefore, the body must rotate about a line through this point.

1.7 ■ "SMALL" MOTIONS

We have been referring to "small" motions of a body. A "small" motion of a body is defined as one that produces an acceptably small shift in the positions of the constraint lines. How much shift is acceptable depends on the specific requirements of a given application. An example will illustrate this.

Figure 1.7.1 depicts a 2D body with an attached arm. It is the intended function of this body to carry the arm through a reciprocating pivotal motion such that the arm will enter and be withdrawn from a hole in a neighboring part without making contact with the sides of the hole.

As shown in Fig. 1.7.2, when the body undergoes 5 degrees of rotation, the constraint lines shift and their point of intersection (and therefore, the location of the instant center of rotation) shifts also.

Nevertheless, the arm clears the sides of the hole as it follows its reciprocating path. This mechanism, therefore, meets its requirements. The shift in the position of the constraint lines is acceptably small for this purpose.

1.8 ■ FUNCTIONAL EQUIVALENCE OF CONSTRAINT PAIRS INTERSECTING AT A GIVEN POINT

We learned in section 1.6 how an X constraint and a Y constraint combine to define an instant center of rotation at the intersection of their constraint lines. Now, let us try to work backwards. We start by identifying the desired location of a rotational degree of freedom and then find the proper position for the two constraints. As usual, we limit our consideration to small motions of the body so that we can ignore the effect of any shift of the position of the instant center resulting from that motion.

Suppose that we design the constraint configuration for a 2D body such that it has a single rotational

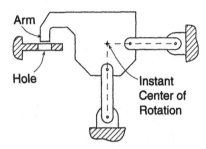

FIGURE 1.7.1

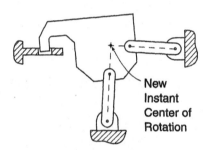

FIGURE 1.7.2

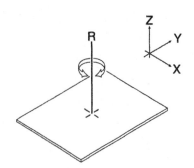

FIGURE 1.8.1

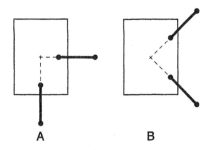

A   B

FIGURE 1.8.2

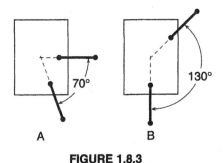

A   B

FIGURE 1.8.3

degree of freedom through its centroid, as indicated in Fig. 1.8.1. First, we know that the constraints must lie in the X–Y plane because this is a 2D problem. Second, we know that we must apply exactly two constraints, because the body is to end up with one degree of freedom. Third, we know that the constraints (or their lines) must intersect at the desired instant center.

However, there is nothing to clue us in about the proper *angle* of the constraint pair with respect to the part. For example, we have a completely arbitrary choice between the geometry of Fig. 1.8.2, A and B.

In fact, we also find that there is no uniquely correct angle *between* the two constraints. For example, the constraint pair shown in Fig. 1.8.3A defines exactly the same instant center as that of Fig. 1.8.3B. It seems that we have a lot of latitude in choosing the location of the constraints as long as their lines intersect at the desired instant center. Of course, we must be careful to avoid *overconstraint*. As discussed in section 1.5, overconstraint occurs when two constraints are acting along the same line. Thus, we must ensure that the angle between the two constraint lines does not approach 0 degrees (180 degrees). When the two constraint lines intersect at a very shallow angle, we approach the condition where the degree of freedom being constrained by one constraint will then be *overconstrained* by the second constraint. This is a lose–lose situation. We lose once when we pay the penalty for overconstraint of one degree of freedom, then we lose again because the orthogonal degree of freedom (which should have been constrained by the second constraint) goes *unconstrained*.

It should, therefore, be clear that in choosing the angle between two constraints whose purpose is to *prevent* translational motion in two orthogonal directions (X and Y), a 90-degree intersection is best. Angles approaching 0 degrees (180 degrees) are the worst and should be avoided.

We can now state the *equivalence* of intersecting **C** pairs:

> Any pair of constraints whose constraint lines intersect at a given point, is functionally equivalent to any other pair in the same plane whose constraint lines intersect at the same point. This is true for small motions and where the angle between the constraints does not approach 0° (180°).

Thus, an intersecting pair of Cs defines a *disk of radial lines*, any two of which (provided the angle between them is not too small) may be selected to equivalently replace the original two Cs.

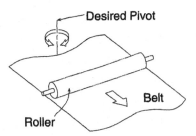

FIGURE 1.9.1

1.9 ■ VIRTUAL PIVOT EXAMPLE

To illustrate how useful the technique of creating a virtual pivot (instant center) can be, consider this design example.

The roller of Fig. 1.9.1 rests on a belt that is being conveyed (by means not shown) in the direction indicated by the arrow. We want to establish a pivot axis for the roller at an "upstream" location as shown. By pivoting the roller from the machine frame at this point, the roller will self-align to the belt in a manner analogous to the way a trailer is towed by an automobile.

Unfortunately, because of space limitations in the machine (of which this apparatus is a part), our design must keep clear of the region upstream of the roller. We must, therefore, establish a virtual pivot at the desired location using constraints that are placed remotely from the pivot.

In the 2D schematic design of Fig. 1.9.2, constraints $C1$ and $C2$ are positioned downstream of the roller, but their lines intersect at the desired upstream pivot axis.

Figure 1.9.3 shows an actual hardware implementation of the schematic design of Fig. 1.9.2. The intersection of the constraint lines of the two links defines a virtual pivot axis for the roller. For the small motions encountered as the roller self-aligns, the position of the virtual pivot does not move substantially.

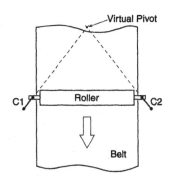

FIGURE 1.9.2

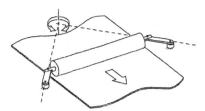

FIGURE 1.9.3

1.10 ■ PARALLEL CONSTRAINTS

At first glance, it would seem that the case of two parallel constraints is categorically different from that of two intersecting constraints. However, it is really just a special case where the constraint lines intersect at a point at infinity.

Consider the diagram of Fig. 1.10.1 where constraint pair $C1$ and $C2$ intersect to define an instant center located at a distance, d, from the body.

Now, imagine the body is permitted to rotate about this instant center a small amount, resulting in

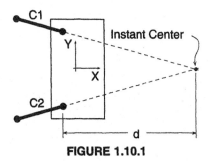

FIGURE 1.10.1

a 1-mm excursion in the Y direction of the body's centroid.

Next consider the effect of reducing the angle between $C1$ and $C2$ to double the distance, d. If the body is again rotated by an amount that results in the same 1-mm Y excursion, we would observe that the body would have rotated through a *smaller* angle.

As we continue to increase distance d, we see that the angle between $C1$ and $C2$ becomes ever smaller and the rotation of the body about its instant center becomes ever smaller to produce the same 1-mm Y motion. Clearly, we can see that as distance d approaches infinity, the constraints $C1$ and $C2$ become parallel, and the angle of rotation about the instant center (located an infinite distance away) needed to produce 1-mm of Y motion becomes zero.

This little experiment allows us to make some important observations.

First:

Parallel lines intersect at infinity.

This is important. When we did our experiment, recall that as we gradually increased distance d, we always maintained the requirement that $C1$ and $C2$ intersect.

The second observation is that we can use rotation about a remote axis to approximate a translation. If the axis is located a great distance away, rotation of the body about this axis (R) is *substantially* equivalent to a pure translation (T).

A T can be equivalently expressed as an R located at infinity.

As a practical matter, a translation can be *approximated* by a remotely positioned R. The R need not be at infinity, nor even a great distance away. Consider, for example, the design of a simple desktop stapler. Although the distance between the stapling head and the hinge is a mere 6 in. or so, the stapling head accomplishes its vertical "translation" quite satisfactorily.

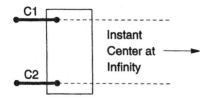

FIGURE 1.10.2

1.11 ■ FUNCTIONAL EQUIVALENCE OF PARALLEL CONSTRAINT PAIRS

It should be appreciated that as a consequence of the principle of the equivalence of intersecting constraint pairs, any pair of constraints, parallel to a given parallel pair of constraints, will intersect at the same point at infinity and therefore be functionally equivalent.

If three different people were asked to apply constraints to the rectangular 2D body of Fig. 1.11.1 in such a way that only the Y translation of the body remains unconstrained, they might give the three solutions shown. Each solution is correct. In fact, there are an infinite number of correct solutions.

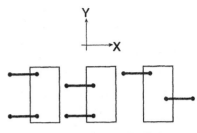

Equivalent Constraint Pairs

FIGURE 1.11.1

> If two lines of an infinite plane of parallel lines represent a pair of Cs applied to a body, then any two parallel lines from that plane can equivalently represent the two Cs applied to the body.

Of course, as always, we must be careful to avoid *overconstraint*, which would occur if the constraints act along the same line (or nearly so). Because the constraints are parallel, it is the *distance* between them, not the angle, that must not be allowed to get too small.

To avoid overconstraint, we must ensure that the distance between the two Cs is not small relative to the dimensions of the body.

1.12 ■ EXACT CONSTRAINT

Up to this point, we have discussed bodies that have one or more degrees of freedom left unconstrained. Let us now look at the special and important case where all three degrees of freedom of a body are constrained.

To start, consider the 2D body of Fig. 1.12.1 constrained in two degrees of freedom by C1 and C2. This body has one rotational degree of freedom, R1, located at the intersection of C1 and C2. This degree of freedom can be constrained by adding a third constraint, C3, whose line is located at a distance, d, from R1. One might think of C3 as "exerting a moment" about R1, where d is the "moment arm." To be effective, moment arm d must not be small relative to the dimensions of the body.

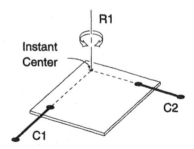

FIGURE 1.12.1

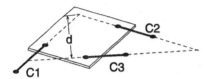

FIGURE 1.12.2

Three constraints have been applied to the body in a way that has reduced the body's number of degrees of freedom to zero. No degree of freedom has been overconstrained, nor has any degree of freedom been left unconstrained. The 2D body is said to be *exactly constrained*.

We could say that the 2D body is "totally" or "completely" constrained. But these terms do not emphasize the fact that each and every one of the body's degrees of freedom has been individually accounted for and constrained, one constraint for each freedom. By using the term exactly constrained, we convey the sense of rigor in our analysis.

The term *exact constraint* also reminds us that more constraint is not necessarily better. (Recall the problems of overconstraint described in section 1.5.)

In general, a good test for determining whether the three constraints applied to a 2D body have been properly placed is to examine the triangle defined by the three constraint lines. Each constraint line needs to have a "good size" moment arm about the instant center of rotation defined by the intersection of the other two constraint lines. Therefore, the perpendicular distance from each side of the triangle to the opposite vertex should not be small relative to the dimensions of the 2D body being constrained.

1.13 ■ CONSTRAINT DEVICES

We use the term constraint device to refer to a mechanical connection whose purpose is to constrain one or more of a body's degrees of freedom. We have already used a few different constraint devices in our models: the link, the contact point, and the bar with articulating end connections. As we proceed, we will learn of many more constraint devices, where the choice of one device or another will be governed by the particular requirements of a given job. But regardless of which constraint devices are chosen, a designer must go through the same thought process of designing the pattern of constraints that will properly constrain each part of a machine, leaving only those degrees of freedom that are desired.

1.14 ■ THE "PIN-AND-HOLE" CONNECTION

The "pin and hole" connection between bodies is a common constraint device intended to provide constraint in both X and Y. However, the pin-and-hole connection is, by definition, an *overconstraint*. Recall from section 1.5 that overconstraint presents us with a choice between (A) loose, sloppy, imprecise parts that go together with "play," (B) parts that bind or do not go together because they are too tight, or (C) expensive parts that fit "perfectly" because of close tolerances or special assembly techniques. In the case of a pin and hole, the pin diameter can be (A) smaller than the hole, (B) larger than the hole, or (C) exactly the same as the hole diameter.

To understand why this connection is an overconstraint, let us look closely at the situation where the pin diameter is smaller than the hole diameter, shown exaggerated in Fig. 1.14.1, A and B. In Fig. 1.14.1A, the right side of the pin makes contact with the right side of the hole. This contact point defines an X constraint. Figure 1.14.1B shows the *left* side of the pin making contact with the *left* side of the hole. This contact point defines *another* X constraint along the same line as the first. Two sets of features (contact points) define two constraints that compete to control the same X degree of freedom. From our definition in section 1.5, this is clearly an overconstraint. The same situation exists in the Y direction. Thus, the pin-in-hole connection is overconstrained in two degrees of freedom, both X and Y.

Does this mean we should *always* avoid using a pin-in-hole connection? No. It has lots of practical uses when only coarse precision of location is required. But when extremely good precision is required, it should definitely be avoided. A far better choice would be the pin-in-"V" connection, shown in Fig. 1.14.2. This connection uses only one set of features (therefore, only one constraint) for each degree of freedom.

Now, let us revisit the link, which uses two pin-and-hole connections, one at each end. Although this served us well as a model for 2D constraints, we now realize that the pin-and-hole connections at the ends are themselves overconstraints. If we intend to use links in an application that requires great precision, we ought to try to improve the design by replacing the round holes with "vees," as suggested in Fig. 1.14.3. Here, a wire spring is used to supply the force needed to nest the pins in their respective vees.

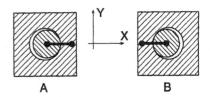

FIGURE 1.14.1

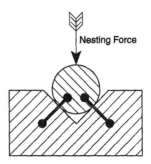

FIGURE 1.14.2

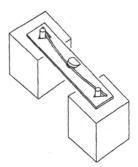

FIGURE 1.14.3

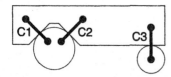

FIGURE 1.15.1

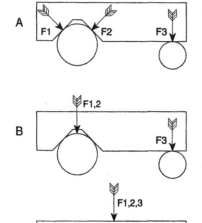

FIGURE 1.15.2

1.15 ■ NESTING FORCES

Recall from section 1.3, when we introduced the contact point as a constraint device, we pointed out that an accompanying nesting force was required to hold the two bodies in contact. Sometimes, it is desirable to have all of a body's constraints provided by contact points. For example, this might be a good way to design the connections to a body that needs to be precisely positioned but also frequently removed and replaced.

The individual nesting forces for each contact point can then be added vectorally to arrive at a single net nesting force needed to maintain contact at all the contact points. Consider the example of Fig. 1.15.1, which depicts a 2D body positioned on two pins.

The individual nesting forces needed for each of the constraints is shown in Fig. 1.15.2A.

In Fig. 1.15.2B, nesting forces $F1$ and $F2$ have been combined by vector addition.

Finally, the net nesting force vector, $F_{1,2,3}$, has been found, which holds the part in its "nest" with uniform force at each contact point. This is the vector sum of F_1, F_2, and F_3, shown in Fig. 1.15.2C.

The nesting force can be applied by various means, as depicted in Fig. 1.15.3.

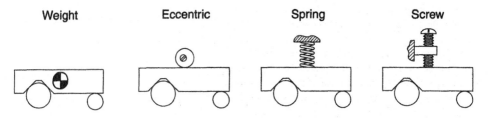

FIGURE 1.15.3

1.16 ■ NESTING FORCE *WINDOW*

In the previous section, we saw how to combine the individual nesting forces at each constraint to a single resultant force. One might wonder how accurately this single net nesting force must be placed to ensure reliable nesting of the part.

There is a window through which the net nesting force must pass in order to achieve reliable nesting. That window can be found by using the following graphical procedure. As an example, we will use the part configuration shown in Fig. 1.15.1. We extend the constraint lines of $C1$, $C2$, and $C3$ to their intersections, as shown in Fig. 1.16.1.

Now imagine for a moment that the part is nested on the large pin, but not quite touching the small pin, as suggested in Fig. 1.16.2. At this instant, constraints $C1$ and $C2$ would be present, but $C3$ would be absent. The part would be constrained to rotate about an axis through the intersection of $C1$ and $C2$. This is the instant center of rotation defined by constraints $C1$ and $C2$. We label it IC 1-2. Because we want the part to become nested against the small pin ($C3$), our nesting force must cause the part to rotate clockwise about IC 1-2.

Next, imagine that our part is temporarily constrained only by $C2$ and $C3$, and $C1$ is absent, as suggested in Fig. 1.16.3. The intersection of constraint lines $C2$ and $C3$ defines IC 2-3, about which the part is constrained to rotate. For the part to make contact at $C1$, the nesting force must cause the part to rotate counterclockwise about IC 2-3.

Finally, imagine that the part is momentarily constrained only by $C1$ and $C3$, and $C2$ is absent. By a similar analysis, we would conclude that in order for the part to make contact at $C2$, the nesting force must cause the part to rotate counterclockwise about IC 1-3.

These three rotational requirements are summarized in Fig. 1.16.4. They must *all* be respected by the net nesting force. The net nesting force must pass each of the three instant centers in a direction that "agrees" with the rotational arrows. This will happen as long as the nesting force passes through a "window," which we will now find by construction.

First, we segment each of the extended constraint lines at the intersection points. Then we draw a thick line on the segments where a force cannot be found to both cross the line *and* cause correct rotation about

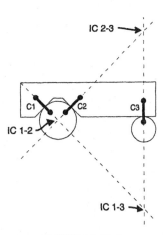

FIGURE 1.16.1

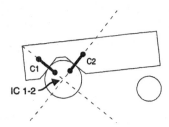

FIGURE 1.16.2

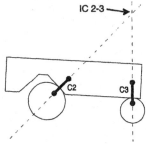

FIGURE 1.16.3

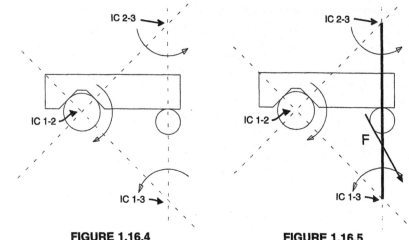

FIGURE 1.16.4   **FIGURE 1.16.5**

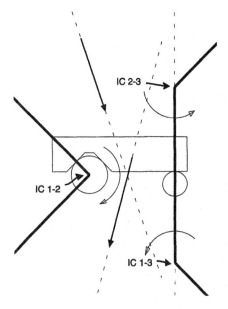

FIGURE 1.16.6

both instant centers on that line. For example, the vertical line of $C3$ has been broken into three segments: an inner segment between IC 2-3 and IC 1-3, an upper segment extending above IC 2-3, and a lower segment extending below IC 1-3. If we imagine a nesting force, F, crossing the inner segment as in Fig. 1.16.5, we will find that it cannot simultaneously satisfy the rotational requirements at both IC 2-3 and IC 1-3. The force F shown produces correct rotation about IC 2-3, but is wrong for IC 1-3. Therefore, the inner segment of the vertical line of $C3$ is drawn as a thick line, which the line of action of the nesting force is *not* permitted to cross.

After completing this analysis for the nine segments of this example, we arrive at the diagram of Fig. 1.16.6. Figure 1.16.6 shows the window through which the net nesting force must pass to achieve nesting of the part *in the absence of friction* at the contact points (constraints). A force applied through the window along either of the two vectors shown will cause the part to be nested properly against the two pins.

Accounting for Friction

Figure 1.16.2 showed our part poised to rotate clockwise about IC 1-2. In order for the part to accomplish this rotation, sliding must occur at the contact points of constraints $C1$ and $C2$. Friction will accompany this sliding. The net nesting force is doing two things: it is providing a nesting force along the constraint lines of $C1$ and $C2$, and it is also trying to cause the part to rotate clockwise about IC 1-2. To produce the desired rotation, friction at $C1$ and $C2$ must be overcome.

We now digress briefly to examine a graphical technique for analyzing friction. This technique will be useful in our analysis of the net nesting force window.

Consider block A resting on surface B, as in Fig. 1.16.7, where the coefficient of friction between A and B is μ. Now imagine doing a test where we apply a force, F, to block A and where we are able to *vary* angle ϕ. Clearly, for angles of ϕ that are small, block A will *not* slide on surface B. Friction will *lock* the block in place. No motion will occur regardless of the magnitude of force F. Only when we apply force F at an angle $\phi = \phi_f$ will the block begin to slide.

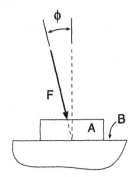

FIGURE 1.16.7

The friction angle, ϕ_f, is that angle at which

$$f = \mu \times N$$

where f is the sliding component of force and N is the normal component, as shown in Fig. 1.16.8.

$$\phi_f = \tan^{-1}\left(\frac{f}{N}\right) = \tan^{-1}\mu$$

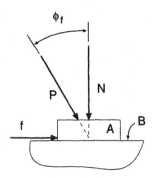

FIGURE 1.16.8

Let us now return to our analysis of the nesting force window and consider the effect of friction on our problem. Suppose we know the value of the coefficient of friction μ between our part and the pins against which it rests. We can find then the friction angle, $\phi_f = \tan^{-1}\mu$. In Fig. 1.16.9, we have drawn the friction angle on each side of the surface normal at the contact point of $C3$.

We next draw the friction angle at *each* contact point and find the *overlap* of these friction angles. Finally, we shade the zones where the friction angles *overlap* and add this to the window found earlier in Fig. 1.16.5. This gives us Fig. 1.16.10, which shows a somewhat smaller window through which the net nesting force vector must pass in order to achieve reliable nesting of our part, accounting for friction at the contact points. The window of Fig. 1.16.5 has been *reduced* in size by the addition of the three shaded *friction zones*. A friction zone surrounds each instant center of rotation. The net nesting force must not pass through any friction zone.

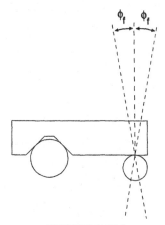

FIGURE 1.16.9

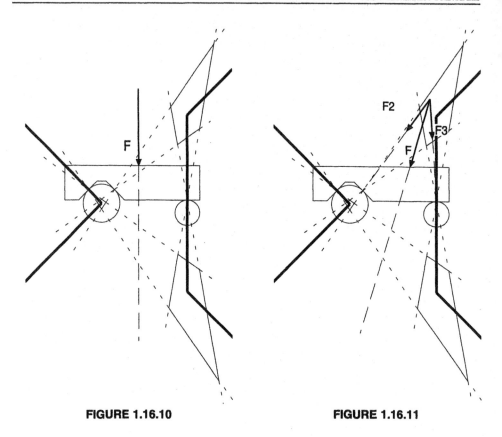

FIGURE 1.16.10   **FIGURE 1.16.11**

Let us examine what would happen if we position our net nesting force so that it passes through a friction zone. Figure 1.16.11 shows force F positioned so that it passes through the friction zone surrounding IC 2-3. Notice that force F can be resolved into two components F_2 and F_3. F_2 is directed at the contact point of $C2$ and lies within the friction angle at $C2$. F_3 is directed at the contact point of $C3$ and lies within the friction angle at $C3$. Because F_2 and F_3 lie within their respective friction angles, no sliding will occur at $C2$ or $C3$. The desired rotation will not occur. The part will not be nested.

1.17 ■ PRECISION AND ACCURACY

In machine design, we frequently encounter the terms "precision" and "accuracy."

Precision (of position), also called repeatability, is the degree to which a part, or feature on a part, will return to exactly the same position time after time. When a machine is designed so that its parts are *exactly constrained*, extraordinary precision is automatically obtained.

For example, suppose we install body A in contact with the lower part B, as suggested by Fig. 1.17.1. Then we measure distance *d*. If we repeatedly remove body A and replace it, each time checking distance *d*, we will be hard-pressed to measure any variation in distance *d*.

Accuracy (of position) is the degree to which the location of a part or feature exactly coincides with its *desired* or intended location. In general, precision is achievable without accuracy, but accuracy is not achievable without precision. By designing *exactly constrained* connections, we can achieve high-precision machines using ordinary, low-cost, inaccurate parts. Then, through the use of adjustments or fixturing techniques, accuracy can be achieved.

For example, say dimension *d* is intended to be 1.000. It might actually turn out to be 1.004 due to dimensional inaccuracies in either or both parts. One can imagine various ways to compensate for dimensional inaccuracies.

In one scheme, shown in Fig. 1.17.2, a gage is used to hold the two parts in position so that dimension *d* is fixtured at 1.000.

Half-round part C is then slid up a ramp on lower part B until it contacts part A. Part C is then bolted securely or glued to part B.

In another scheme, shown in Fig. 1.17.3, dimension *d* is first measured. Then, based on this measurement, a shim of an appropriate thickness is selected and placed between half-round part C and lower part B. The half-round part and shim are then bolted securely to the lower part.

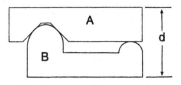

FIGURE 1.17.1

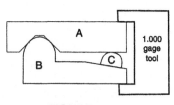

FIGURE 1.17.2

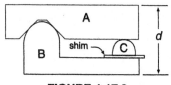

FIGURE 1.17.3

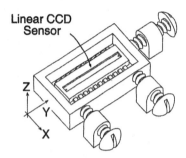

Linear CCD Sensor

FIGURE 1.18.1

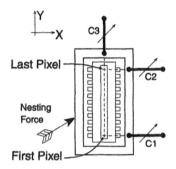

Y
X
C3
Last Pixel
C2
Nesting Force
C1
First Pixel

FIGURE 1.18.2

A   B

FIGURE 1.18.3

1.18 ■ THE SCREW: AN ADJUSTABLE CONSTRAINT DEVICE

One of the advantages of designing *exact constraint* connections is being able to achieve a high degree of precision with relatively inexpensive, inaccurate parts. However, to achieve *accuracy* in an assembly of inaccurate parts, we must resort to accurate assembly fixtures or adjustments. Adjustments can be readily designed into a mechanism by using ordinary machine screws. The end of the screw is used as a contact point constraint device. Then, by turning the screw, the position of the object being constrained can be adjusted, one degree of freedom at a time. By judicious selection of the positions of these constraints, the adjustments can be made noninteractive.

Figure 1.18.1 shows a linear charge coupled device (CCD) sensor containing a row of light-sensitive pixels (picture elements). Each pixel is 10 μm square and the row of pixels is 25 mm long.

The position of the end pixels must be accurately aligned with respect to a reference image, both in X and Y. The three screws are used to make this adjustment. Figure 1.18.2 shows the situation diagrammatically, with the adjustable constraints shown symbolically.

In the nominal position, the constraint lines of C1 and C3 intersect at the first pixel, so adjustment of C2 causes no motion at the first pixel, only an X motion at the last pixel. Similarly, because the constraint lines of C2 and C3 intersect at the last pixel, adjustment of C1 causes no motion of the last pixel, only an X motion of the first pixel. Finally, adjustment of C3 causes only a pure Y motion of the entire row of pixels.

Once the adjustments are made, they can be locked by using either a jam nut or a clamp nut, as shown in Fig. 1.18.3.

1.19 ■ DESIGNING FOR INSENSITIVITY TO THERMAL EXPANSION

Suppose we need to mount an apparatus (a body) such that a particularly critical feature remains in a fixed position irrespective of any thermal expansion of the body induced by temperature shifts.

Figure 1.19.1 shows such a body with a bore whose axis must not move.

The effect of thermal expansion of the body (shown exaggerated by the dotted outline of the enlarged part) causes every point to move radially in proportion to its distance from a given point (origin). We designate the center of the bore as the origin, and then we orient all constraint surfaces radially with respect to the bore center. The result is that the position of the bore is insensitive to thermal expansion.

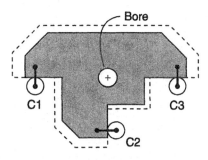

FIGURE 1.19.1

1.20 ■ 2D CONSTRAINT PATTERN CHART

Table 1.20 summarizes the various patterns of orthogonal constraints that can be directly applied between a 2D body and a reference body (not shown).

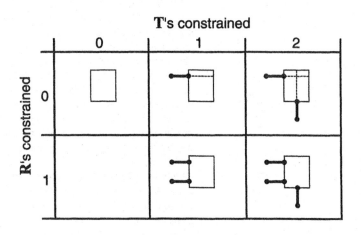

TABLE 1.20

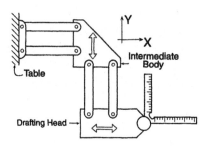

FIGURE 1.21.1

1.21 ■ CASCADING

The 2D constraint pattern chart, which summarizes the constraint patterns directly applicable to a 2D body, shows no pattern that will constrain the rotation of a 2D body while leaving both translations free. To achieve rotational constraint without translational constraint, we use a technique known as *cascading*. This requires the introduction of an intermediate body. In cascading, the body of interest is connected by a first set of constraints to the intermediate body, which, in turn, is connected by a second set of constraints to the reference body (ground). With cascaded connections, the degrees of freedom allowed by the first connection are *added* to those allowed by the second connection. A drafting machine is a good example of cascading. The connection between the head and the intermediate body allows only X freedom. The connection between the intermediate body and the table allows only Y freedom. Therefore, the head relative to the table is free only in X and Y. It is constrained in θ_z.

1.22 ■ ROTATIONAL CONSTRAINT

Cascading is not the only means available for achieving a pure rotational constraint without translational constraint. In Fig. 1.22.1, a body with two pulleys attached is constrained rotationally by a single cable, whose ends are both attached to a reference body. A nesting torque must be applied, as shown. The body is then free in both X and Y. This is exactly the connection used on some drafting boards to keep the vertically traversing straightedge from rotating. The nesting torque is applied by an opposed cable.

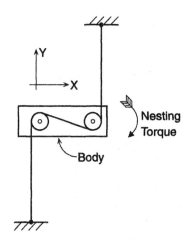

FIGURE 1.22.1

1.23 ■ THE INTERSECTION OF Rs AND Cs

Recall from section 1.3 the statement on the effect of a constraint:

Points on the object along the constraint line can move only at right angles to the constraint line, not along it.

This led us in section 1.6 to realize that a 2D body's rotational degree of freedom (**R**) must be located somewhere along (intersect) the line of any constraint applied to the body. We can state this more broadly as follows:

The axes of a body's rotational degrees of freedom
(**R**s) will each intersect all constraints (**C**s)
applied to the body.

As discussed in section 1.10, "intersect" is again
understood to include the condition of "being parallel
to" where intersection occurs at a point at infinity.
This not only explains why instant centers are located as they are at the intersection of two constraints,
but we will see that it can tell us where *all* of a body's
rotational degrees of freedom are located.

The discovery that a body's **R**s intersect all
applied **C**s is the foundation of Exact Constraint
Design. Throughout the remaining chapters of this
book, we build on this foundation and, in the process,
make many more interesting and useful discoveries
about the design of machines.

1.24 ■ TS EQUIVALENTLY EXPRESSED AS RS AT INFINITY

We are now poised to find a body's degrees of freedom (**R**s) given an arbitrary pattern of applied **C**s.
But what if one or more of the degrees of freedom
turns out to be translational degrees of freedom (**T**s)?

No problem. A **T** is expressed as an **R** at infinity.
Earlier, in section 1.10, we found that a translational
degree of freedom **T** can be equivalently represented
as an **R** at infinity. Thus we can now consider an
alternative way of expressing the degrees of freedom
of the 2D body, which we depicted in Fig. 1.1.1 as
two **T**s and one **R**.

Each of the **T**s can be equivalently expressed as
an **R**, located at an infinite distance away, as shown
in Fig. 1.24.1. **R**1 is shown at an arbitrary position on
the body.

Figures 1.24.1 and 1.1.1 are equivalent represen-
tations of an unconstrained 2D body's degrees of
freedom. In Chapter 3, we further develop the relationships that exist between patterns of lines in space
representing the **C**s applied to a body and their
resulting **R**s.

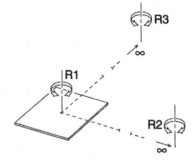

FIGURE 1.24.1

CHAPTER SUMMARY

In this chapter we defined constraints (Cs) and freedoms (Rs). We discovered that when a constraint is applied to a body, the constraint is represented by a line in space, along which no motion is permitted. Overconstraint occurs when two constraints are located along the same line (or nearly so). A mechanical connection between two 2D bodies can be represented as a pattern of constraint lines in the 2D plane. When the connection between two bodies consists of two constraints, we discovered the existence of patterns of "equivalent" lines. The pattern depends on the geometrical relation between the constraint lines. If the lines intersect at a finite location, then a radial "disk" of lines is defined, any two of which can be equivalently substituted for the original pair of C lines. If the lines are parallel (the lines intersect at infinity), then an infinite planar array of parallel lines is defined, any two of which can be equivalently substituted for the original pair of C lines. When a pattern of two constraints is applied to a 2D body, it will have one rotational degree of freedom (R) located at the intersection of the constraint lines, perpendicular to the 2D plane. In the case where the two Cs are parallel (intersecting at infinity), the resulting R is located at the intersection at infinity and this is equivalent to a pure translational degree of freedom.

When constraints are applied *directly* between two bodies, the resulting connection is represented by that pattern of C lines. When two bodies are connected in *cascade*, such that one body is connected through one or more intermediate bodies to the other body, the resulting connection is represented by all Rs of each cascaded connection.

CHAPTER *2*

Three-Dimensional Constraint Devices

In our study of 2D connections between objects, we used some simple constraint devices, such as the contact point and the link with pivoting end connections. We learned that these constraint devices are functionally equivalent for small motions and that for the purpose of our analysis, it does not matter which constraint device we decide to use. It is only important to know whether and where constraints are applied.

The same is true for 3D connections between objects. As we learn more about 3D connections, we will discover a larger array of constraint devices. Some of these devices provide single constraints. Some provide multiple constraints in a pattern.

Nevertheless, we continue to represent each constraint using the now familiar constraint symbol (Fig. 1.3.4). Our analysis is concerned with the locations in space of these constraints, not with the specific constraint device(s) being used. Each constraint is represented as a line in space.

Having said all of this, there are still reasons why a designer might prefer one constraint device over another. Those reasons might have to do with the intrinsic stiffness of the connection, whether or not it can be loaded bidirectionally, ease of assembly and disassembly, range of motion allowed, cost, and so on. A designer's experience will often guide him or her in this selection.

In this chapter, various constraint devices are presented along with some comments about their application.

A   B

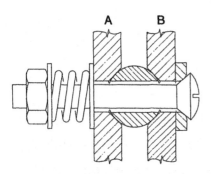

FIGURE 2.1.1

2.1 ■ BALL-AND-SOCKET JOINTS

The ball-and-socket joint is a frequently encountered multiple constraint device. Figure 2.1.1 shows a version of the ball-and-socket joint that includes its own means for application of nesting force.

Constraint devices using a ball-and-socket joint are common in machines and they are even found in nature. For example, the human hip is a ball-and-socket joint. This type of joint provides three constraints intersecting at the center of the ball. It permits rotation about three axes that intersect at the center of the ball.

The connection between a towing vehicle and trailer is another example of a ball-and-socket connection. In this case, the joint is overconstrained because the socket engages the ball over an area that is greater than a hemisphere. The penalty for this overconstraint is that the joint has some looseness and play. The benefit is that it provides the joint with the capability to carry large loads in any direction, up, down, right, left, forward, or backward. For this application, the omnidirectional load-carrying capability is important; the small amount of play in the joint is not a serious problem.

However, there are applications (e.g., in the design of precision instruments) where looseness and play are not tolerable. For such applications, care must be taken to ensure that the connection between the ball and socket is not overconstrained. A ball in a trihedral socket avoids overconstraint by presenting three orthogonal surfaces against which the ball is nested. (The shape of a trihedral socket is that which is formed by shoving the corner of a cube into a mound of clay.) A single nesting force, applied along the line between the ball centerline and the apex of the trihedral socket, ensures that all three contact points remain engaged. This connection, although it is "kinematically correct," can be improved. It can be made stiffer and stronger by using a technique that I call *curvature matching*.

2.2 ■ CURVATURE MATCHING

Consider the situation of a steel ball loaded against the flat surface of a body. Nesting force P is applied along the constraint line. Indentation occurs at the "point" of contact between the ball and the body, resulting in high contact stress and reduced stiffness. Both of these problems can be reduced by causing the radius of curvature of the contacting surfaces to become more nearly matched.

The configuration of the ball nesting in a trihedral socket can also be improved (in two respects) by going to a *conical*-shaped socket. First, the socket is easier to machine. Second, the engagement between the ball and the socket goes from three points (small areas) to a line (a ring-shaped area). Borrowing the terminology we use to describe our globe, if we define the polar axis of the ball to be coincident with the axis of the conical hole, then we have achieved curvature matching along a line of latitude (but not along meridional lines). This substantially improves the stiffness and reduces the stress at the ball–socket interface without imposing tight tolerances on the form of the socket. In other words, the fit of the ball into the socket is not affected by small variations in the diameter of the ball or angle of the conical socket.

To further improve stiffness and reduce stress at the ball–socket interface, we must address the match between the curvatures of the ball and socket along the meridional lines. Obviously, a spherical-shaped socket that exactly fits the ball provides the ultimate result that is achievable by curvature matching. Unfortunately, this type of fit may be difficult to achieve in practice. Small differences between the radius of curvature of the ball and that of the socket can spoil the fit. To avoid this problem while still achieving some measure of curvature matching along meridional lines, the radius of curvature of the socket in the meridional direction should be made *slightly larger* than that of the ball. Figure 2.2.3 shows, in section view and with exaggerated curvature, a ball nested in such a socket. The cross-sectional shape of this socket resembles that of a Gothic arch. In actual practice, the meridional radius of curvature of the socket may approach the radius of the ball more closely. It is important to realize that the curvatures of the ball and socket *need not match exactly* to achieve higher stiffness

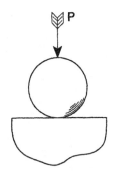

FIGURE 2.2.1

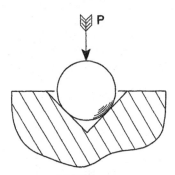

FIGURE 2.2.2

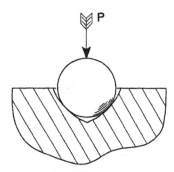

FIGURE 2.2.3

and lower stress in a connection. The benefits of curvature matching are realized when we design the radius of a part to be *more nearly like* that of its mating surface.

2.3 ■ SINGLE-CONSTRAINT DEVICES

The simplest constraint device, the round-ended contact point, applies a single constraint irrespective of small amounts of misalignment. The line of the C goes through the point of contact, perpendicular to the surface tangent at the point of contact. Unfortunately, the point of contact of the ball on the flat surface can result in high stress and reduced stiffness. This situation can be improved by using the principle of curvature matching. An intermediate body can be installed between A and B such that the curvatures of the contacting surfaces of A and B are both matched. This intermediate body is the C-clamp foot.

In Fig. 2.3.2, a circular line of contact is achieved against the ball. An area contact is achieved against the flat surface. The connection is now cascaded. The C-clamp foot is the intermediate body. This configuration is stiffer than the situation shown in Fig.2.3.1.

Figure 2.3.3 shows a variation on the C-clamp foot, which includes a means for applying its own nesting force. Notice that small amounts of misalignment between A and B will be accommodated.

With each of the contraint devices of Figs. 2.3.1, 2.3.2, and 2.3.3, any motion of the body in the plane perpendicular to the constraint line will result in sliding at the contact surfaces of the constraint device. The designer must be careful to ensure that the friction at those sliding surfaces does not interfere with whatever freedom of motion is intended. This may require the designer to specify a low nesting force, or the designer may prefer to select a different constraint device, such as a bar with ball-and-socket end connections. The one shown in Fig. 2.3.4 provides a single constraint along the line of centers of the two balls while providing very free motion perpendicular to the constraint line.

A slender bar or wire connected between two bodies can be used to apply a single constraint along its axis. Its slenderness allows it to bend to accommodate small rotations and translations normal to its axis while still providing high stiffness *along* its axis. This device is called a wire flexure and is discussed in greater detail in Chapter 3.

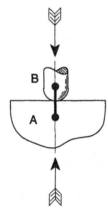

FIGURE 2.3.1

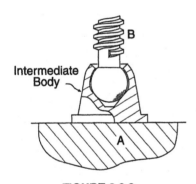

FIGURE 2.3.2

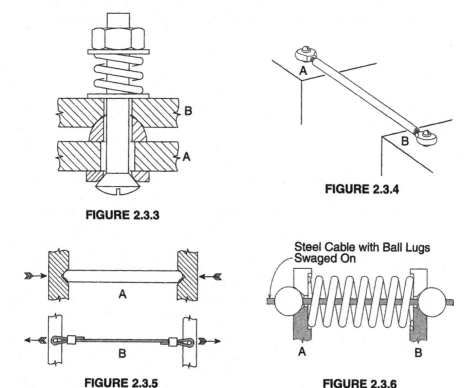

FIGURE 2.3.3

FIGURE 2.3.4

FIGURE 2.3.5

FIGURE 2.3.6

If it is not required to support bidirectional loads, then one of the constraint devices of Fig. 2.3.5 can be used. Figure 2.3.5A is a round-ended bar nested in a dimple in each body. Figure 2.3.5B is a wire cable. The bar will support compression loads only; the cable will support only tension loads.

Figure 2.3.6 shows an improvement to the cable idea in which a coil spring has been added. This constraint device can support both tension loads and compression loads up to the magnitude of the spring force. Also, it accommodates misalignment easily.

Of course, the device of Fig. 2.3.6 does not need to use a flexible cable. Because the ball lugs are free to articulate, the cable could be replaced by a rigid steel bar.

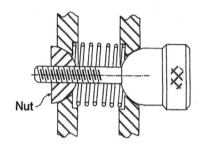

Nut

FIGURE 2.4.1

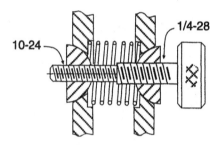

10-24

1/4-28

FIGURE 2.4.2

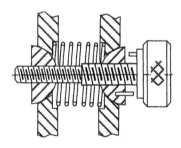

FIGURE 2.4.3

2.4 ■ ADJUSTABLE CONSTRAINT DEVICES

If we need an *adjustable* constraint, we could replace the bar with a screw. One lug becomes the nut and the other is the screw head.

When designing very fine adjustments, we often use fine pitch screws. An 80-tpi (threads per inch) screw, which is about as fine as is practical, advances about .0015 in. with each ⅛ turn. If a still finer pitch is needed, a differential screw can be used. A differential screw has two threaded sections, one having a slightly different pitch than the other. The effective pitch is the difference between the two pitches. For example, a 24-tpi screw has a thread pitch of 0.0417 in. The pitch of a 28-tpi screw is 0.0357 in. The difference is 0.006 in., which is equivalent to a 168-tpi screw.

In Fig. 2.4.2, the screw is now threaded with a 10-24 thread on one end and a ¼-28 thread on the other. Each of the lugs is a nut. Friction between each nut and its seat keeps the nut from spinning when the screw is turned.

The problem with using a differential screw with such a fine effective pitch is that it has a very limited range. For example, if the 24-tpi and 28-tpi screws are long enough to accommodate ten turns of adjustment (about 0.400 in. of axial motion allowed to each nut), the differential screw will produce only ¹⁄₁₆ in. of motion. A differential screw capable of accommodating a large adjustment range would end up needing to be very long and would require a great deal of patience on the part of the person using it!

This problem has been neatly solved by David Kittell of Stamford, CT, who devised this differential screw. The compact design provides both a large range *and* a fine effective pitch by operating in two modes. In the *coarse* mode, one of the nuts (the finer one) is forced to spin with the screw, thus preventing the differential effect. In this mode, when only the 10-24 screw is active, a large range of adjustment can be obtained. In the *fine* mode, both screws are active, producing the differential effect. The device is in the fine mode when the pin in the ¼-28 nut is *disengaged* from the pin extending from the underside of the screw head. This occurs for just one turn of the screw head. In operation, the screw is turned (in coarse mode) until just *past* the proper setting, then it is backed off a little, then turned in again (less than one turn); this time in fine mode.

This device is effectively an adjustable length bar with ball-and-socket ends. The nuts, having a hemispherical shape, nest in conical holes in the two bodies being connected. The included angle of each conical hole is 60 degrees. This ensures that friction will keep the nuts from spinning in their seats when they should not. A shallow counterbore in each of the bodies provides a seat for a coil spring, which applies an axial nesting force.

2.5 ■ WHEELS

The wheel provides a single Z constraint while permitting a large (infinite) range of travel. The wheel of Fig. 2.5.1 has infinite and effortless range of travel in the X direction; however, Y direction motion is accompanied by skidding. If the wheel is supporting a heavy load, and if friction is high, it will not be adequately free to accommodate Y motion. For such a situation, a self-aligning castered wheel might be preferable.

The castered wheel of Fig. 2.5.2 will rotate about the vertical spindle to accommodate motion anywhere in the X–Y plane. The behavior of the castered wheel is discussed in greater detail in Chapter 8.

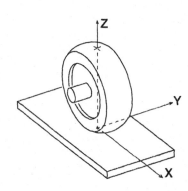

FIGURE 2.5.1

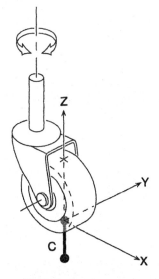

FIGURE 2.5.2

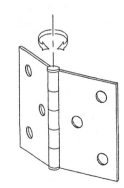

FIGURE 2.6.1

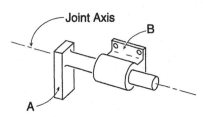

FIGURE 2.6.2

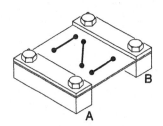

FIGURE 2.6.3

2.6 ■ MULTIPLE CONSTRAINT DEVICES

The hinge, sketched in Fig. 2.6.1, is another multiple constraint device that is commonly found in machinery. The hinge provides five constraints. The one freedom permitted is rotation about the hinge axis.

It the hinge is constructed to permit a range of axial motion, too, we would have the device pictured in Fig. 2.6.2. This device imposes four constraints oriented radially with respect to the joint axis. Two freedoms are permitted: one to translate along the joint axis, and one to rotate about the joint axis.

If we made the shape of the contacting surface between A and B to be some cross-sectional shape other than circular, we would have a "spline" joint. Rotation about the joint axis would be constrained. Only translation along the joint axis would be permitted. Kinematicians refer to the spline joint as the "prismatic" joint.

The sheet flexure, shown in Fig. 2.6.3 connecting between two bodies, behaves somewhat like the wire flexure. It is thin enough to bend and accommodate three **R**s that lie in the flexure plane and yet it is still able to apply three **C**s, which also lie in the flexure plane. The sheet flexure is explored in great detail in Chapter 4.

The last constraint device covered here is the (preloaded) ball bearing. It is by no means last in importance or frequency of use.

A ball bearing will normally have a small amount of clearance between the balls and races. You can detect this clearance by wiggling the inner race relative to the outer race while holding the bearing in your hands.

An axial preload force applied between the inner and outer races, as shown in Fig. 2.6.4, serves as a nesting force. With the nesting force in place, there are *n* obliquely oriented constraints (one for each of *n* balls) between the inner and outer races. The angle of these constraints is referred to as the "contact angle" of the bearing. They all intersect at a common point on the bearing axis.

Obviously, a bearing with more than three balls will have more than three constraints, and thus be overconstrained. For this reason, the balls and races must be manufactured accurately. It is clear from Fig. 2.6.4 that a shaft mounted in a single bearing still has three degrees of freedom. Those three degrees of freedom are **R**s through the common intersection point of the *n* constraints. It is interesting to note that we saw this same pattern of three **C**s and three **R**s intersecting at one common point in the ball-and-socket joint.

CHAPTER SUMMARY

In this chapter, we explored the three-dimensional constraint patterns of several familiar 3D connections as we expanded our "catalog" of constraint devices from the single-constraint devices used in Chapter 1.

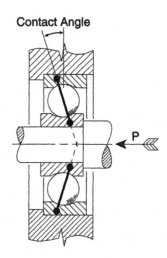

FIGURE 2.6.4

CHAPTER 3

Three-Dimensional Connections Between Objects

The 2D models of Chapter 1 introduced many important concepts. However, we must now step away from this simplification and consider the whole three-dimensional situation. In this chapter, we analyze connections between objects in 3D. Of course, it is well known that a free body has six degrees of freedom: three independent translational degrees of freedom (often referred to as X, Y, and Z) and three independent rotational degrees of freedom (often referred to as θ_x, θ_y, and θ_z).

What is not well known, however, is the exact manner in which mechanical connections to a 3D object effect a reduction in the number of its degrees of freedom from six (in its free state) to some smaller number (as a result of the connection).

For example, consider the situation of Fig. 3.0 in which an object is connected to a base body by three arms (links). Clearly, these three arms have constrained the object and have thus reduced the number of degrees of freedom the object has relative to the base.

It is not very clear, however, just what degrees of freedom, if any, remain between the object and the base. It turns out that there are some fairly simple, but not well-known constraint pattern analysis techniques that enable us to analyze connections such as the one in Fig. 3.0 and correctly deduce the object's degree(s) of freedom. This chapter reveals techniques for solving problems such as this.

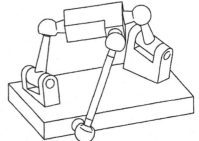

FIGURE 3.0

35

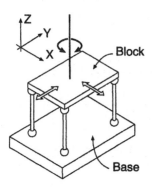

FIGURE 3.1.1

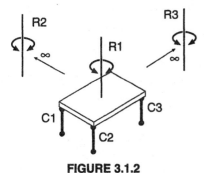

FIGURE 3.1.2

3.1 ■ 3D MODEL OF THE 2D CASE: CONSTRAINT DIAGRAM

The model shown in Fig. 3.1.1 was constructed using a block of wood and three thin wooden sticks. Hot-melt glue was used to attach the sticks to the block and to the base. Each of the sticks is a constraint device which, for small motions, behaves like a bar with ball-and-socket joints at each end.

If you were to build such a model, you would notice the block moves quite freely in three degrees of freedom, X, Y, and θ_z, but that it is quite rigidly attached to the base in the remaining three degrees of freedom, Z, θ_x, and θ_y. Each of the bars applies a *constraint* between the block and the base. The three bars together have constrained three of the block's degrees of freedom. By applying this pattern of three constraints to the block, three of the block's degrees of freedom have been *removed*.

The block's three remaining degrees of freedom (X, Y, and θ_z) are precisely the same as the three degrees of freedom of the unconstrained 2D model of Chapter 1.

We can draw a *constraint diagram* representation of this model, as in Fig. 3.1.2. In this diagram each bar is represented by a constraint symbol ($\mathbf{C}$). Each of the body's degrees of freedom is expressed as a rotational degree of freedom ($\mathbf{R}$).

As we learned previously, the body's translational degrees of freedom (X and Y) can be equivalently expressed as $\mathbf{R}$s located at infinity. Notice that the block has three $\mathbf{C}$s and three $\mathbf{R}$s and that they are all parallel, *as they must be*. In section 1.23, we discovered that a body's $\mathbf{R}$s will intersect the applied $\mathbf{C}$s. In Fig. 3.1.2, in order for *each* $\mathbf{R}$ to intersect *each* $\mathbf{C}$, the $\mathbf{R}$s must be parallel to the $\mathbf{C}$s, intersecting them at infinity.

We will make extensive use of constraint diagrams as we explore the many possible mechanical connections that occur between objects. A constraint diagram allows us to visualize easily both the pattern of $\mathbf{C}$ lines imposed on a body by a given mechanical connection, and the resulting pattern of $\mathbf{R}$ lines representing the body's rotational degrees of freedom.

3.2 ■ RELATIONSHIP BETWEEN Rs AND Cs

Now let us consider the effect of adding a fourth bar, $C4$, to our model, as in Fig. 3.2.1. This is precisely the same constraint condition as shown in Fig. 1.3.2, except that now we have explicitly shown constraints C_1, C_2, and C_3. The result of this connection, of course, is exactly as we found it to be in section 1.3. The application of one additional C results in the loss of ˚one R. The sum of the number of Cs applied (without overconstraint) plus the number of remaining degrees of freedom (Rs) equals six.

The body now has two degrees of freedom, shown in Fig. 3.2.2 as two Rs. $R1$ represents the body's θ_z degree of freedom. $R2$ (at infinity) represents the body's translational degree of freedom.

Both $R1$ and $R2$ are shown intersecting the line of $C4$. We know this must be true. Recall from the definition of a constraint that a constraint prevents motion along the constraint line and allows motion only perpendicular to the constraint line. This requires that any remaining rotational degrees of freedom (Rs) must intersect the constraint line. This condition must be true for *each* and *every* constraint applied. Examining Fig. 3.2.2, we see that this does indeed hold true. $R1$ and $R2$ intersect $C4$ and also intersect $C1$, $C2$, and $C3$ at infinity, as they are all parallel lines.

This relationship between the pattern of lines representing an object's Rs and the pattern of lines representing the Cs applied to the object is very important. It allows us to find an object's Rs given a mechanical connection that imposes some pattern of Cs. Whenever we know a pattern of Cs applied to an object, we can find the resulting pattern of Rs from this rule:

> When a pattern of *n* constraints is applied between an object and a reference body (without overconstraint), the object will have 6-*n* rotational degrees of freedom (Rs), and each R will intersect each C.

The idea of "intersection" of a C line with an R line is a very mutual sort of thing. That is, if a C intersects an R, then it is just as true that the R intersects the C. Carrying this idea a little further, if we

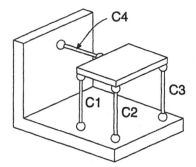

FIGURE 3.2.1

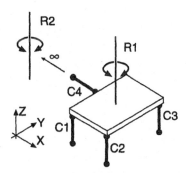

FIGURE 3.2.2

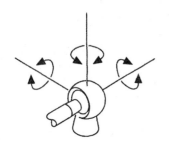

FIGURE 3.2.3

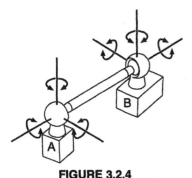

FIGURE 3.2.4

can find a pattern of **R** lines, given a pattern of **C** lines, then we ought to be able to just as well find a pattern of **C** lines given a pattern of **R** lines. Given one pattern, we ought to be able to find the "complementary" pattern. Given the **C**s, we can find the **R**s. Given the **R**s, we can find the **C**s.

Let us look at an example. Consider the ball-and-socket connection shown in Fig. 3.2.3. It is well known that a ball-and-socket connection constrains X, Y, and Z translational motion but permits θ_x, θ_y, and θ_z rotation.

The constraint diagram for this connection reveals a pattern of three **R**s intersecting at the center of the ball. The complementary pattern is three **C**s also intersecting at the center of the ball. Every line of one pattern intersects every line of the complementary pattern.

Now consider the connection between bodies A and B shown in Fig. 3.2.4, consisting of a rigid bar connected at its ends by ball-and-socket joints. This is a cascaded connection between bodies A and B, where the bar is an intermediate body. Because the connection from A to B is cascaded, we must *add* the *freedoms* of each connection. The total pattern of freedoms consists of three **R**s intersecting at the center of one ball and three more **R**s intersecting at the center of the second ball. The total number of **R**s is six. If we try to find the complementary pattern of **C** lines, we seem to get a contradiction. Because there are six **R** lines, the arithmetic says we should get 0 **C** lines. On the other hand, there is one line that intersects every one of the **R** lines. The line joining the center of the two balls intersects every line in the **R** pattern. Intuitively, we know that this line represents the single constraint imposed by this connection. The sixth **R** is redundant, the **R** line along the bar axis (joining the two ball centers) is provided twice, once by each ball-and-socket joint. Therefore, we have *underconstraint* in one degree of freedom. The effect of this is to allow the bar to rotate about this axis.

Returning now to the rule for finding **R**s from a pattern of **C**s, recall that when we started with a pattern of **C**s, we required that the pattern not be *overconstrained*. In other words, the pattern of **C**s must not contain any redundant lines. When we apply the *converse* of the rule (starting with a pattern of **R**s) we again find that in order for the arithmetic to work (number of Rs + number of Cs = 6), there must be no redundant lines in the pattern.

Thus, we can now restate this rule in a more general manner so that we can use it to go *bidirectionally* between patterns of Cs and Rs. We call this the **Rule of Complementary Patterns**.

When a pattern of C lines is imposed between two objects, there is a resulting and complementary pattern of R lines that exists between the two objects. Given one or the other of these patterns containing n lines without redundancy, the complementary pattern will contain $6-n$ lines. Furthermore, every line of one pattern will intersect every line of the complementary pattern.

We can use the Rule of Complementary Patterns to find the resulting pattern of Rs given an arbitrary pattern of Cs. Let us go through some examples.

In Fig. 3.2.5, we have added $C5$ to our model. Using the Rule of Complementary Patterns, we know that the pattern of five Cs will result in a single R. Furthermore, we know that this single R must intersect all five Cs. There is only one possible location for this R. It passes through the intersection of $C4$ and $C5$ and is parallel to $C1$, $C2$, and $C3$, intersecting them at infinity. There is no other line in space that intersects all five Cs.

Figure 3.2.6 shows another object constrained by five constraints. $C1$, $C2$, and $C3$ intersect at point A. $C4$ and $C5$ intersect at point B. Our rule tells us that the body will have exactly one (= 6 – 5) R. In order for this R to intersect every one of the five Cs, it must be the line joining A and B.

In each of the two previous examples, we had a pattern of five Cs in which three of them intersected at one point and the remaining two intersected at a second point. The resulting R was the line joining the two intersection points. This is the only line that can be found that intersects all five Cs.

Another way to find the R in Fig. 3.2.6 is to notice that all of the Cs lie in two planes. $C1$, $C2$, and $C4$ lie in a vertical plane. The R line that we are looking for must also lie in that plane (since it must intersect $C1$, $C2$, and $C4$). $C3$ and $C5$ define a second plane (which splits the block diagonally). Any R line that lies in this second plane will intersect $C3$ and $C5$. The intersection of the two planes is the

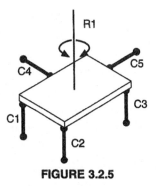

FIGURE 3.2.5

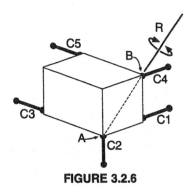

FIGURE 3.2.6

unique location for the **R**. This is the only line in space that is in both planes. Because it lies in both planes, it intersects all five **C**s.

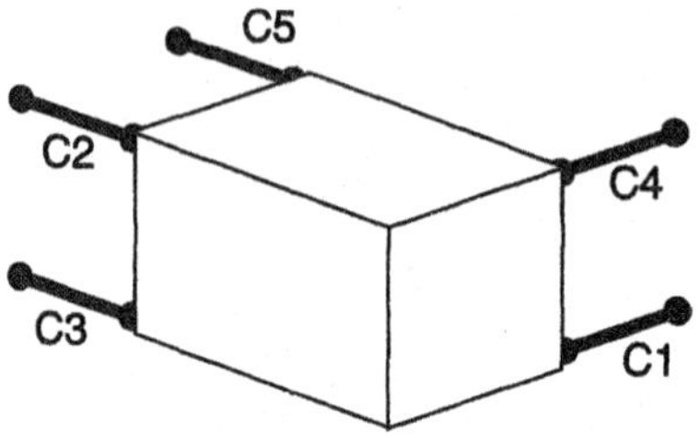

3.3 ■ T EQUIVALENT TO R AT INFINITY

Figure 3.3.1 shows an object constrained by a pattern of five **C**s lying in two parallel horizontal planes. To find the object's single **R** we might look for the intersection of the two planes. We can find the intersection of the two planes just like we found the intersection of the two lines in section 1.10. The intersection of the two planes would be a tangent line on a horizontal circle of infinite radius, as suggested by the diagram of Fig. 3.3.2, where the object is located at the center of the circle.

The object has just one degree of freedom. It is a straight vertical translation **T**. This is the equivalent representation of an **R** located tangentially anywhere on the horizontal circle of infinite radius.

FIGURE 3.3.1

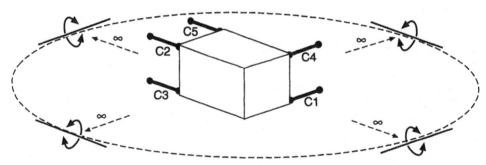

FIGURE 3.3.2

Some readers may have used their intuition to recognize that the degree of freedom of the object of Fig. 3.3.1 is a vertical translation. After all, there are no vertical bars, just horizontal ones. So there is clearly no constraint to prevent vertical motion. To such a reader, the approach of first finding the **C**s in two parallel planes, next finding the **R** as the intersection of those two planes at infinity, then finally replacing the **R** at infinity with its equivalent vertical translation must have all seemed quite tedious and abstract.

I must agree. This is a case where constraint pattern analysis led us to the correct solution in a fairly roundabout way. However, I can assure you that there will be times when you will face problems too complex to yield to an intuitive solution, but these problems will yield easily to a solution using constraint pattern analysis techniques.

3.4 ■ INTERSECTING PAIR OF **R**s: DISK OF RADIAL LINES

Consider the constraint pattern of Fig. 3.4.1 where an object is constrained by a pattern of four **C**s. Three of the **C**s are coplanar and lie in the plane of the top face of the object. The fourth **C** attaches to the object at the lower right corner. The Rule of Complementary Patterns tells us the object will be left with two **R**s, each intersecting every one of the four **C**s. Those two **R**s will be two lines from a disk of lines whose center is intersected by **C**4 and which lies in the plane of **C**1, **C**2, and **C**3. The body's two **R**s are not *uniquely* determined. Any two lines from this disk will intersect every one of the four **C**s.

Regardless of which two lines we might arbitrarily select from the disk to represent the body's two **R**s, the two we specify will be correct.

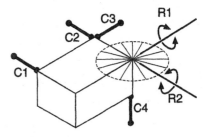

FIGURE 3.4.1

Any pair of intersecting rotational degrees of freedom (**R**s) is equivalent to any other pair intersecting at the same point and lying in the same plane. This holds true for small motions.

This is exactly analogous to the rule of equivalence of intersecting **C**s that we discovered in section 1.8. There, we discovered that an intersecting pair of **C**s defined a planar disk of radial lines where any two could be equivalently substituted for the original pair, as long as the angle between the two lines was not too small. When the angle becomes too small, we found that we approached the condition of overconstraint.

In the present case, where we have two intersecting **R**s from a planar disk of lines, we might wonder whether we will also need to be cautious about selecting two lines separated by just a very small angle. Let us look at a practical example of a real hardware design problem that has two intersecting **R**s. We will solve the problem using our knowledge of the planar disk of equivalent radial lines.

Consider the roller, illustrated in Fig. 3.4.2, that is required to have two degrees of freedom. In addition

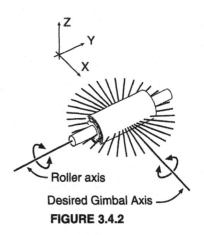

Roller axis

Desired Gimbal Axis

FIGURE 3.4.2

FIGURE 3.4.3

to being free to roll (θ_y), it must also gimbal a small amount about the X-axis (θ_x). Unfortunately, because of space restrictions, no hardware can actually be allowed to occupy the space along the X-axis.

Because the two required **R**s intersect at the roller center, they define a planar disk of radial lines (in the X–Y plane), any two of which can be used to substitute for θ_x and θ_y. If we are precluded from designing a pivoting yoke that provides θ_x freedom, we can select another, more convenient line from the disk and achieve the same result.

Recall in section 1.21 that if we design a "cascaded" connection, we add the degrees of freedom of each connection. Our problem is solved by designing a cascaded mechanism whereby the roller shaft is connected to the machine frame so that it pivots about an oblique axis, **R**2. The bearings supporting the roller on its shaft define **R**1. The two **R**s intersect at the roller center and lie in the X–Y plane. Thus, the roller has two intersecting **R**s. For small rotations about the oblique axis, this produces precisely the same effect as if the roller were actually gimbaled about the X-axis.

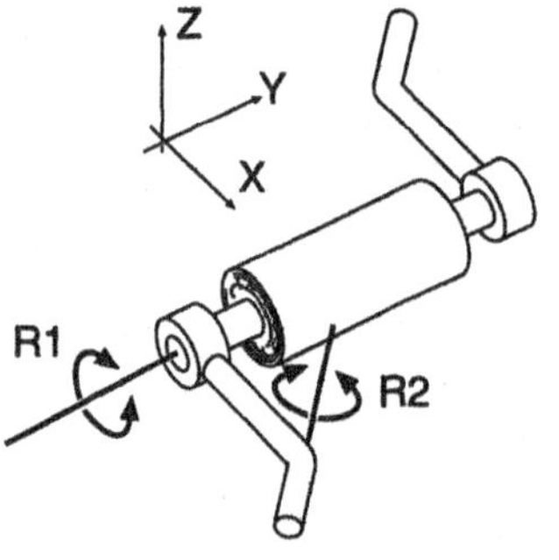

Now consider the size of the angle between **R**1 and **R**2. Because our roller is required to gimbal (θ_x) only a few degrees (perhaps to self-align to some other part of the machine), we consider it to be a "small motion." We can easily achieve a "substantially pure" θ_x motion of a few degrees even though **R**2 is located at a 45-degree angle from **R**1 (and from θ_x). In fact, we could have made the angle between **R**1 and **R**2 even smaller than 45 degrees.

However, if we think about what would happen as the angle between **R**1 and **R**2 approaches 0 degrees, it should be clear that we would then have *underconstraint* of the roller shaft in θ_y, while our θ_x degree of freedom would vanish.

So, we find that the situation we have with two intersecting **R**s is *exactly* analogous to the case of two intersecting **C**s.

An intersecting pair of **C**s or **R**s defines a disk of radial lines any two of which (provided the angle between them is not too small) may be selected to equivalently replace the original pair (for small motions).

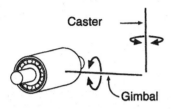

FIGURE 3.4.4

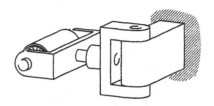

FIGURE 3.4.5

Another example illustrates this principle. The body shown in Fig. 3.4.4 must be mounted so that it is free to caster and gimbal a small amount about the axes shown.

The obvious solution, shown in Fig. 3.4.5, cannot be implemented, because other machine components are already using that space.

The solution is to first show the disk of radial lines defined by the desired caster and gimbal axes, then design a cascaded mechanism that provides $R1$ and $R2$, each of which is a member line of that disk.

The disk is shown in Fig. 3.4.6. It is defined to be in the same vertical plane as the caster and gimbal axes, with its center at the intersection of the two axes.

For small motions, the mechanism of Fig. 3.4.7 accomplishes the desired caster and gimbal motions shown in Fig. 3.4.4, even though it provides rotation about two *other* lines from the disk.

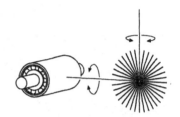

FIGURE 3.4.6

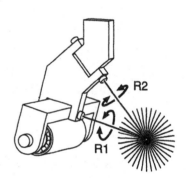

FIGURE 3.4.7

3.5 ■ TWO PARALLEL Rs: A PLANE OF PARALLEL LINES

Let us return to the 3D model of the 2D case with two degrees of freedom discussed in section 3.2. Recall the model of Fig. 3.2.1 showing $C1$, $C2$, and $C3$, all parallel and therefore intersecting at infinity, and horizontal constraint $C4$. We previously found the body to have two degrees of freedom, $R1$ and $R2$, where $R2$ was located at infinity and represented the body's translational degree of freedom. Note that $R1$ and $R2$ are just two parallel lines from an infinite set of parallel lines, all lying in one plane, that intersect $C4$ and are parallel to lines $C1$, $C2$, and $C3$. The Rule of Complementary Patterns tells us that given this pattern of four Cs, the

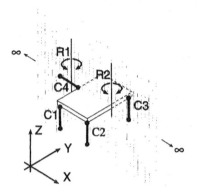

FIGURE 3.5.1

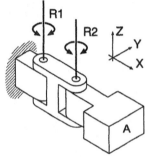

FIGURE 3.5.2

complementary pattern of $\mathbf{R}$s can be any two lines from this infinite plane of parallel lines. We are not *required* to specify any two particular lines. We are not required, for example, to state that one of the object's degrees of freedom is horizontal translation in Y.

We would be perfectly correct to say that the body's two degrees of freedom are $\mathbf{R}1$ and $\mathbf{R}2$, both located on or near the body, as shown in Fig. 3.5.1. $\mathbf{R}1$ and $\mathbf{R}2$ have been *arbitrarily* chosen from the plane of parallel lines defined by constraint pattern $\mathbf{C}_1$–$\mathbf{C}_4$.

To demonstrate this, imagine building a connection to a block, A, that consists of two cascaded parallel hinges, as shown in Fig. 3.5.2. Clearly, A would have two rotational degrees of freedom, $\mathbf{R}1$ and $\mathbf{R}2$. Now, imagine that we conceal this connection inside a "black box," and allow only the block to be revealed outside of the box. We then invite someone to grasp the block and test it to discern its degrees of freedom for small motions. It is likely our tester would tell us that the object has two degrees of freedom: θ_z rotation and Y translation.

It is very unlikely that our tester, by exercising the object through small motions, would be able to detect that the object is, in fact, constrained by two hinges, nor be able to discern their locations.

The object's two degrees of freedom, when exercised simultaneously, combine to give the object the rotational degree of freedom of any of the infinite number of parallel lines in the plane of parallel lines defined by $\mathbf{R}1$ and $\mathbf{R}2$. For example, if the object rotates clockwise a small amount about $\mathbf{R}1$ and simultaneously rotates an equal and opposite amount about $\mathbf{R}2$, the object will have accomplished a translation without rotation. We know this is equivalent to a rotation about an $\mathbf{R}$ at infinity. By simultaneously rotating about $\mathbf{R}1$ and $\mathbf{R}2$ in *various proportions*, the object can accomplish an effective rotation about any of the infinite number of parallel lines in the plane of parallel lines defined by $\mathbf{R}1$ and $\mathbf{R}2$.

We are thus led to the following truth:

> If two lines of an infinite plane of parallel lines represent two $\mathbf{R}$s of a body, then any two parallel lines from that plane can equivalently represent the two $\mathbf{R}$s of that body.

This conclusion, of course, is merely a corollary of the situation observed in section 3.4, where an intersecting pair of **R**s defined a planar disk of equivalent lines. The only difference is that in the present case the intersection point is at infinity. Also recall that we discovered the same truth about parallel **C**s in section 1.11.

3.6 ■ REDUNDANT LINES: OVERCONSTRAINT AND UNDERCONSTRAINT

It is important to be able to recognize when a pattern of lines contains redundant lines. A pattern of **C**s with redundant lines is overconstraint. A pattern of **R**s with redundant lines is underconstraint. When we use the Rule of Complementary Patterns to find a complementary pattern, we must be aware of any redundant lines in our starting pattern.

We have learned enough now to be able to find redundancies in various patterns. The simplest redundancy is when two lines are collinear. We learned in section 1.5 that the two collinear **C**s result in overconstraint. Similarly, two collinear **R**s cause an underconstraint. Figure 3.6.1 suggests a cascaded connection consisting of two collinear hinge axes. The body seems to have two **R**s, but, in fact, the **R**s are redundant. The object really has only one **R**. The second **R** results in an independent degree of freedom for the intermediate body.

We see that regardless of whether the lines are **C**s or **R**s, two collinear lines are redundant.

Next, consider two lines (either **R**s or **C**s) that lie in the same plane. Because they are coplanar, they must intersect. (If they are parallel, they intersect at infinity.) Because this pair of lines defines a disk of radial lines any two of which are equivalent to the initial pair, a third line added to the pattern (of the same type, **R** or **C**) lying in the same plane would be redundant if it intersects the same point. (Once the initial pair of lines has defined the disk of radial lines, it would be redundant for a third line to also be one of these radial lines.)

Continuing this line of reasoning, a nonredundant pattern of three coplanar lines can be shown to be equivalent to a disk of radial lines at *any* point in the plane plus a third line not a member of the disk. Thus, every line in the plane is already represented by the initial pattern of three coplanar lines. A fourth line added to the pattern would be redundant if it lies anywhere in the same plane as the first three.

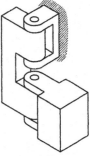

FIGURE 3.6.1

In a similar fashion, if we started with a non-coplanar pattern of three lines intersecting at a point, we would discover that every line intersecting that point is a member of a radial "star" of lines defined by the initial three lines. Thus, a fourth line added to the pattern would be redundant if it intersected the same point. This holds true, of course, even for parallel lines, whose intersection point is at infinity.

This may seem all quite obvious. At least it was to James Clerk Maxwell over 100 years ago when he parenthetically summarized all of the above patterns of overconstraint in a scientific paper on the subject of the proper design of scientific instruments.[4] Referring to a body's constraints, he said: "No two coincide; no three are in one plane, and either meet in a point or are parallel; no four are in one plane, or meet in a point, or are parallel, or, more generally, belong to the same system of generators of an hyperboloid of one sheet. The conditions for five . . . and for six are more complicated."

The truth, of course, is that these constraint patterns are not well known. They have become obscured by time. So we will dust them off and expand the definition to include patterns of **R**s.

Whereas, Maxwell listed various patterns of lines that represent overconstraint, we will call these redundant patterns. If the lines are **C**s, then indeed we have overconstraint. If the lines are **R**s, then we have underconstraint.

Once we have a pattern of lines that is not redundant, we can then use the Rule of Complementary Patterns to find the complementary pattern of lines.

3.7 ■ COMPOUND CONNECTIONS

Often, parts of a machine are connected in patterns that include both direct and cascaded connections in combination. These are called compound connections. Analyzing compound mechanical connections is somewhat analogous to simplifying an electrical resistance network having both parallel and series connections. For example, suppose we were given the electrical resistance network shown in Fig. 3.7.1, consisting of various series (cascaded) and parallel (direct) connections between nodes A, B, and C.

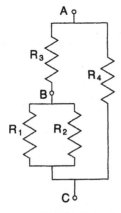

FIGURE 3.7.1

[4] Maxwell, J. C. *General considerations concerning scientific apparatus.* In *The Scientific Papers of J. C. Maxwell* (vol. II.) W. D. Niven (ed.). Cambridge University Press, London, 1890, pp. 507–508.

Suppose we want to find the equivalent connection between A and C, R_{A-C}. There are two parallel *paths* of connection between A and C: the left path, through R_3, R_2, and R_1, and the right path, through R_4. To simplify the electrical resistance network and solve for R_{A-C}, we must start on the "inside" and work our way out.

We first combine R_1 and R_2 to find R_{B-C}. Because R_1 and R_2 are in parallel, we would add "conductances."

$$\frac{1}{R_{B-C}} = \frac{1}{R_1} + \frac{1}{R_2}$$

Next, we combine R_3 and R_{B-C} in *series* to get R_L, the resistance of the left path.

$$R_L = R_3 + R_{B-C}$$

Finally, we combine R_L and R_4 in parallel to get R_{A-C}.

$$\frac{1}{R_{A-C}} = \frac{1}{R_L} + \frac{1}{R_4}$$

In summary, we are able to simplify the electrical resistance network by recognizing parallel and series connections and adding conductances or resistances, as appropriate.

When we try to resolve complex mechanical connections, we must do a similar thing; but with mechanical connections, we must recognize *direct* and *cascaded* connections and add **C**s or **R**s, as appropriate.

Suppose we are given the analogous mechanical apparatus shown in Fig. 3.7.4, and are asked to find the constraint pattern (or the complementary **R** pattern) that exists between bodies A and C.

We can borrow the schematic diagram from the electrical network analysis to describe our mechanical connections. The *mechanical* schematic diagram is shown in Fig. 3.7.5.

First, we observe two connection paths between A and C: the top path, consisting of three direct constraints, and the bottom path, which is cascaded through arm B. Just as with the electrical network, we

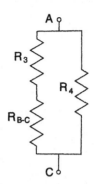

FIGURE 3.7.2

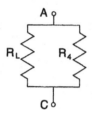

FIGURE 3.7.3

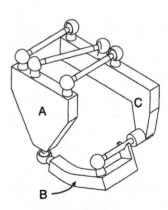

FIGURE 3.7.4

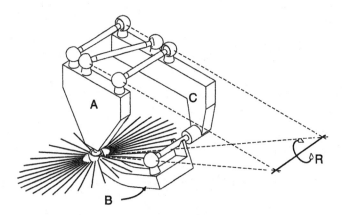

FIGURE 3.7.5

find that our solution must begin at the inside and proceed outward.

We start with the B–C connection, a direct (parallel) connection of a ball-and-socket (three Cs) and a bar in a "V" (two Cs). This combines to five Cs. The complementary pattern is one R through the center of the ball and through the intersection of the two Cs of the "V." Now we can solve the lower path. Because it is cascaded, we add Rs. The A–B connection has three Rs through the center of the ball. Add that to the R just found in the B–C connection for a total of four Rs. The complementary pattern of Cs (for the lower path) is two radial lines from a planar disk of radial lines through the center of the A–B ball-and-socket, and in the plane of the R of the B–C connection. These two Cs of the lower path combine with the three coplanar Cs of the upper path to a total of five Cs. The complementary pattern is one R, located at the intersection of the two planes, the plane of the three Cs of the upper path, and the plane of the two Cs of the lower path.

FIGURE 3.7.6

3.8 ■ VECTOR ADDITION OF COUPLED Rs

Consider body A, which is connected by a hinge to an intermediate body, which in turn is connected by a second hinge to "ground," as shown in Fig. 3.8.1. This is a cascaded connection, consisting of two 5-constraint (one degree of freedom) connections in series. The effect of this is to add the freedoms permitted by each connection. In this case, the freedom permitted by each connection is a single **R** defined by each hinge axis. Thus body A has two **R**s, **R**1 and **R**2.

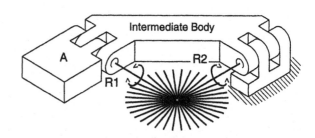

FIGURE 3.8.1

Because **R**1 and **R**2 intersect, we know they define a planar disk of lines, any two of which could be considered to represent body A's two degrees of freedom. If constraint **C** is applied to body A, as in Fig. 3.8.2, we immediately realize that body A now has just one degree of freedom, and that degree of freedom is the **R** represented by the radial line that is intersected by **C**.

Let us take a closer look at what is actually happening at the two hinges when body A rotates a small amount about

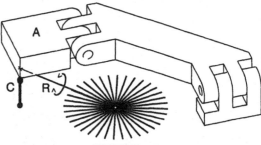

FIGURE 3.8.2

oblique axis, **R**. To be clear about the direction of rotation (cw or ccw), we express each **R** as a vector with an arrowhead symbol to denote direction (+).

Using the standard right-hand rule convention, the direction of rotation is indicated by the fingers of the right hand when the thumb is pointed along the positive direction of the vector.

Now, allowing body A to undergo a small positive rotation about **R**, we can see that rotation about each of the two hinge axes, **R**1 and **R**2, is occurring simultaneously. In fact, if we carefully measure the magnitude of those rotations about **R**1 and **R**2, we discover that they occur in specific proportions, resulting in no motion along constraint **C**.

$$a \times \mathbf{R}1 = b \times \mathbf{R}2$$

We can see from Fig. 3.8.3 that this is the physical requirement imposed by constraint **C**. In order for points on body A along constraint **C** to have zero motion along **C**, any rotation about **R**1 *must* be accompanied by a proportionate rotation about **R**2 such that the two rotations cancel each other out at the location of constraint **C**, which is positioned a distance a from **R**1 and a distance b from **R**2. Rotations **R**1 and **R**2 are said to be "coupled." They are not independent degrees of freedom as they were before the application of constraint **C**.

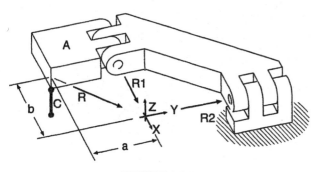

FIGURE 3.8.3

After constraint **C** is applied, body A has just one degree of freedom, **R**. This degree of freedom is equivalent to the simultaneous combination of small rotations **R**1 and **R**2 in the exact proportion:

$$\frac{\mathbf{R}1}{\mathbf{R}2} = \frac{b}{a}$$

These two rotations are the vector components of **R** along the directions of **R**1 and **R**2:

> Coupled rotations about **R**1 and **R**2 combine vectorally to produce resultant **R**.

3.9 ■ HELICAL DEGREE OF FREEDOM (SCREW)

Now consider the situation shown in Fig. 3.9.1, where body A again has two **R**s resulting from a cascaded connection, but this time the **R**s do not intersect as they did in the previous example. Since R_x and R_y do not intersect, they do not define a planar disk of radial lines.

Therefore, when we apply constraint **C** to the body, as in Fig. 3.9.2, to discover what is the body's single remaining degree of freedom, we do not find a quick and easy answer.

To find the body's single remaining degree of freedom, we must find the proportionate values of coupled vectors **R**x and **R**y that will result in zero motion along constraint **C**. We then combine those two vectors to determine their resultant. This will be the body's single remaining degree of freedom.

Before the application of constraint **C**, body A has two degrees of freedom, **R**x and **R**y, which do not intersect but are parallel with the X and Y axes, respectively. Then constraint **C** is applied, parallel to the Z-axis.

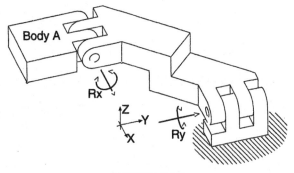

FIGURE 3.9.1

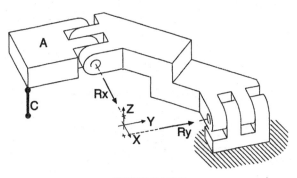

FIGURE 3.9.2

The perpendicular distance between **R**x and **C** is *a*. The perpendicular distance between **R**y and **C** is *b*. After **C** is applied, body A will be incapable of Z direction motion along **C**. Body A will have one remaining degree of freedom which will be the resultant of coupled vectors **R**x and **R**y in the specific proportion: $a \times \mathbf{R}x = b \times \mathbf{R}y$. When small rotations about **R**x and **R**y occur in this specific proportion, no motion along **C** results.

Now, we combine these two coupled vectors **R**x and **R**y to find their resultant. To add them vectorally, we first "slide" them to a common plane where they will intersect each other. Thus we find equivalent vectors in a common plane at an intermediate Z-position. Vector **R**x can be replaced by **R**x' positioned a dis-

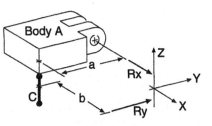

FIGURE 3.9.3

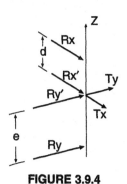

FIGURE 3.9.4

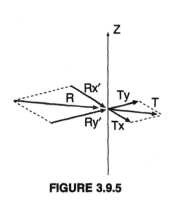

FIGURE 3.9.5

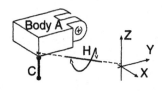

FIGURE 3.9.6

tance d along the Z-axis, coupled with orthogonal translation $Ty = d \times Rx$. Vector Rx' is equal in magnitude and parallel to vector Rx. Vector Ry is replaced by Ry' positioned a distance e along the Z-axis coupled with orthogonal translation $Tx = e \times Ry$. Vector Ry' is equal in magnitude and parallel to vector Ry.

In the new, intermediate plane, vectors Rx' and Ry' will combine to a resultant vector R. Vectors Tx and Ty combine to a resultant vector T. T and R are coupled, just as Rx and Ry were coupled. By judiciously choosing the "correct" position along the Z-axis for the intermediate plane, resultants R and T will be collinear. When coupled vectors R and T are collinear, they combine to produce a pure helical degree of freedom, or screw.[5]

Resultant vectors R and T will be collinear when:

$$\frac{Ty}{Tx} = \frac{Ry}{Rx}$$

but

$$\frac{Ry}{Rx} = \frac{a}{b}$$

and

$$\frac{Ty}{Tx} = \frac{d \times Rx}{e \times Ry}$$

This leaves us with

$$\frac{d}{e} = \frac{a^2}{b^2}$$

Thus, body A's single degree of freedom is a helical degree of freedom, H, whose position is shown in Fig. 3.9.6.

If you were to build this model and test it to find its single degree of freedom, you would discover that you could rotate the body about the H axis, but the rotation would be accompanied by a translation along the H axis.

[5] Ball, R. S. *The Theory of Screws: A Study in the Dynamics of a Rigid Body*. University Press, Dublin, 1876.

It is not possible to get the rotation to occur without the translation. They are *coupled* and therefore represent *one* degree of freedom.

3.10 ■ THE CYLINDROID

We are now able to find the position of body A's single helical degree of freedom resulting from the application of a Z constraint at any arbitrary X–Y position. Let us position that **C** at various points around the circle:

$$x^2 + y^2 = r^2$$

to find all the locations of the resulting **H**. The family of lines is shown in Fig. 3.10.1, all intersecting the Z-axis. These lines are generators of a "ruled" surface known as a cylindroid.[6]

A good way to understand the shape of a cylindroid is to visualize its intersection with the surface of the circular cylinder, $x^2 + y^2 = r^2$, whose axis is the Z-axis. (This cylinder is the surface swept out by constraint **C**.)

If we could mark the surface of the cylinder along the intersection with the cylindroid, then slit the cylindrical surface along one of its generators and lay it out flat, the curve marked would be two periods of a sine wave. The peak-to-peak amplitude of this sine wave would be the same irrespective of the diameter of cylinder chosen. This amplitude represents the length of the cylindroid.

Note the similarity between the apparatus of Fig. 3.9.1, which was used to discover the cylindroid, and the apparatus of Fig. 3.8.1, which defined a planar disk of lines.

Now note the similarity between the planar disk of lines and the cylindroid. Both are "ruled" surfaces of straight, radial lines, perpendicular to the surface axis. The cylindroid resembles the planar disk except that it has been distorted as if it were made of rubber and lines $\mathbf{R}_x$ and $\mathbf{R}_y$ have been pulled apart along the Z axis. Just as the planar disk represents the infinite family of lines, any two of which can be considered to equivalently represent **R**1 and **R**2 prior to the application of constraint **C** in Fig. 3.8.2, so also does the cylindroid represent such an infinite family of equivalent lines. The difference is that in the case of

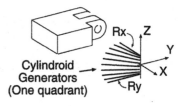

Cylindroid Generators (One quadrant)

FIGURE 3.10.1

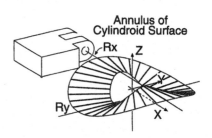

Annulus of Cylindroid Surface

FIGURE 3.10.2

---

[6] Phillips, J. *Freedom in Machinery: Vol. 2: Screw Theory Exemplified.* Cambridge University. Press, 1985.

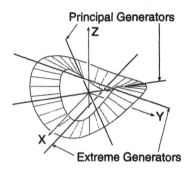

Principal Generators

Extreme Generators

FIGURE 3.10.3

the cylindroid, the generators of the surface are $\mathbf{H}$s, not $\mathbf{R}$s, and the pitch (the magnitude of the coupled translation) of those $\mathbf{H}$s depends on the position of the line on the cylindroid.

Any two $\mathbf{R}$s in space, $\mathbf{R}1$ and $\mathbf{R}2$, will always uniquely define and reside on a cylindroid. That cylindroid will have an axis that is the mutual perpendicular of the two $\mathbf{R}$s and it will have a center plane located half way between the two $\mathbf{R}$s, perpendicular to the axis. The cylindroid will always have two orthogonal, intersecting generators that are contained in the center plane. These are called the *principal generators*. When viewed along its axis, a cylindroid's principal generators will always bisect the angle between the two *defining generators* ($\mathbf{R}1$ and $\mathbf{R}2$). The two generators that are located at the maximum distance from the center plane are called the *extreme generators*. These are the generators that intersect the peaks and troughs of the sine wave. The distance between them is the length of the cylindroid. When viewed along its axis, a cylindroid's extreme generators appear as an orthogonal pair, which is rotated 45 degrees from the pair of principal generators.

In the present case, $\mathbf{R}x$ and $\mathbf{R}y$ are the extreme generators of their cylindroid, since the angle between them is 90 degrees when viewed along their mutual perpendicular, the cylindroid axis (also the Z-axis). All of the other generators of the cylindroid are $\mathbf{H}$s (screws). Any two lines from this cylindroid can be considered to equivalently represent body A's two degrees of freedom, $\mathbf{R}x$ and $\mathbf{R}y$.

3.11 ■ EXAMPLE

Let us now return to the mechanism illustrated in Fig. 3.0 to find the degree(s) of freedom that exists between the upper body and the base. Figure 3.11.1 shows the same mechanism again, but from a slightly different view.

Body A is connected to the base by two arms and a bar. In combination, these represent a "compound connection." We therefore use the technique of section 3.7 to diagram this. Figure 3.11.2 shows the *mechanical schematic diagram* of the mechanism of Fig. 3.11.1.

We start at the inside and work our way out. We first examine the cascaded connection through one of the arms. Each arm has a hinge at one end and a ball-and-socket connection at the other end. Adding the three **R**s of the ball-and-socket joint to the single **R** of the hinge, we get a total of four **R**s, as shown in Fig. 3.11.3.

We next use the Rule of Complementary Patterns to find the pattern of **C**s that is complementary to this pattern of four **R**s. We know there will be two **C**s and they will each intersect every one of the four **R**s. The complementary pattern of two **C**s resides in the planar disk of lines, whose center is the center of the ball, and whose plane contains the hinge axis. Two lines from the disk, shown in Fig. 3.11.4, represent the two **C**s imposed by the arm.

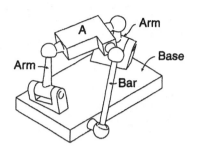

FIGURE 3.11.1

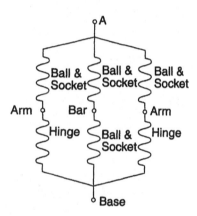

FIGURE 3.11.2

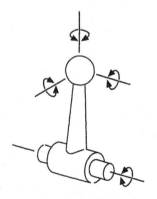

FIGURE 3.11.3

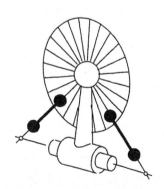

FIGURE 3.11.4

We can now show the disks for each of the two arms in the context of the complete mechanism, as in Fig. 3.11.5. There is a total of five **C**s between the base and the body A: two **C**s from each arm and a single **C** from the bar with ball-and-socket ends.

For the purpose of this analysis, let us ignore the single **C** from the bar for the present time and consider the mechanism as if it had only the two arms connecting body A and the base. We will add the bar again later. Figure 3.11.6 shows the mechanism *without* the bar.

For the mechanism of Fig. 3.11.6, we can use the Rule of Complementary Patterns to find the two **R**s that are complementary to the four **C**s represented by the two disks. One **R** is the line joining the centers of the two disks. The other **R** is the line at the intersection of the planes of the two disks. These two **R**s are shown in Fig. 3.11.7.

R1 and **R**2 define a cylindroidal surface, any two generators of which equivalently represent the body's two degrees of freedom. We will now construct the cylindroid from **R**1 and **R**2, the two defining generators.

First, we find the line that is mutually perpendicular to **R**1 and **R**2. This line is the cylindroid axis. To help visualize the shape of the cylindroid surface, imagine a round cylindrical surface of arbitrary

FIGURE 3.11.5

FIGURE 3.11.6

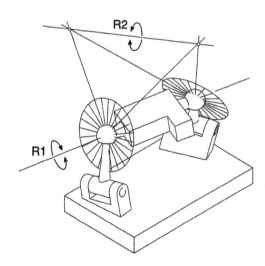

FIGURE 3.11.7

radius, whose axis is coincident with the cylindroid axis, as shown in Fig. 3.11.8.

Now imagine drawing two periods of a sine wave around the perimeter of this cylindrical surface such that the sine wave intersects **R**1 and **R**2 and the zero of the sine wave is located half way between **R**1 and **R**2, as shown in Fig. 3.11.9.

The cylindroid surface is generated by a radial line, perpendicular to the axis, as it traces this sine wave. This surface is shown in Fig. 3.11.10. (Actually, the surface shown is just an *annular section* of the cylindroid.)

When viewed along the axis, the cylindroid appears as a disk and the ruled lines on the surface of the cylindroid appear as radial lines of the disk. For our analysis, body A's two degrees of freedom are represented by two lines from the cylindroid. These two degrees of freedom will be **H**s.

We now replace the bar, as shown in Fig. 3.11.11. The constraint imposed by the bar removes one of body A's two degrees of freedom. To find the single remaining degree of freedom of body A, we use the method developed in Section 3.8.

We must find the perpendicular distance between constraint **C** and each of **R**1 and **R**2. In this example, the perpendicular distance between **C** and **R**1 is

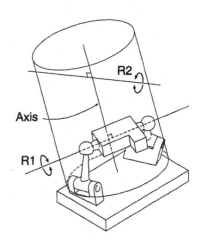

FIGURE 3.11.8

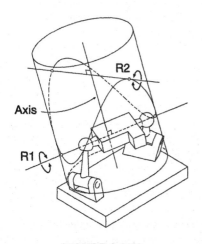

FIGURE 3.11.9

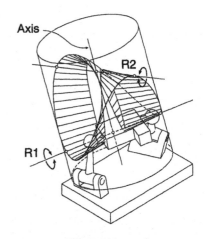

FIGURE 3.11.10

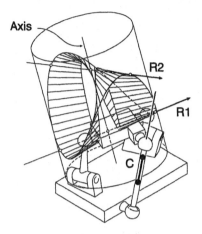

FIGURE 3.11.11

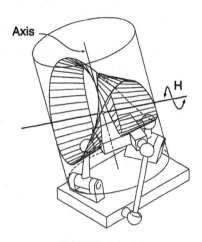

FIGURE 3.11.12

the same as the perpendicular distance between **C** and **R**2. As a result, **R**1 and **R**2 are coupled in equal proportions. **R**1 and **R**2 combine vectorally to a resultant that is located in an intermediate plane half way between **R**1 and **R**2. The resultant vector **H** is at the intersection of this intermediate plane and the cylindroid, as shown in Fig. 3.11.12. This happens to be, by coincidence, one of the principal generators of the cylindroid.

3.12 ■ SUMMARY: SYMMETRY OF RS AND CS

In this chapter, we have extended the fundamental ideas presented in Chapter 1. The constraint diagram was introduced as a tool for visualizing the spatial positions of patterns of lines. The idea of expressing a body's degrees of freedom as six lines in space, each representing a rotational degree of freedom (**R**), was an important step leading to the discovery of the Rule of Complementary Patterns. This rule relates a pattern of **C** lines to its complementary pattern of **R** lines and is an extremely powerful tool that can be used for analyzing existing machines or synthesizing new ones.

We have learned that a pattern of six or fewer lines in space can be used to represent the mechanical connection between two bodies. That pattern might represent the freedoms of a body or it might represent its constraints. If we are given the constraint pattern imposed on a body, we can find the body's freedoms. If, on the other hand, we start with a pattern of freedoms that we intend for a body, we can determine the constraint pattern that, when applied to the body, will result in the desired degrees of freedom. The pattern of lines representing a body's freedoms (**R**s) is complementary to the pattern of lines representing the body's constraints (**C**s). If we know one pattern, we can find the other. Regardless of whether a pattern of lines represents a body's **C**s or its **R**s, the complementary pattern is found in exactly the same way.

We have thus discovered a very basic symmetry that exists *between* **R**s and **C**s:

> Given a pattern of n lines without redundancy, the complementary pattern will contain 6-n lines and every line of one pattern will intersect every line of the complementary pattern.

We have also discovered that within a pattern of lines, regardless of whether those lines represent **R**s or **C**s, certain transformations are possible:

> If a pattern of lines (**R**s or **C**s) contains two lines that intersect, then that intersecting pair defines a radial disk of equivalent lines, any two of which can be equivalently substituted for the original pair.

Because we have defined "intersection" to include parallel lines, where intersection is at infinity, we also have this corollary:

> If a pattern of lines (**R**s or **C**s) contains two parallel lines, then that parallel pair defines a plane of parallel lines, any two of which can be equivalently substituted for the original pair.

Both of these transformations (the radial disk of lines and the plane of parallel lines) carry the limitation that lines must be chosen such that they do not approach the condition of redundancy (overconstraint for **C**s, underconstraint for **R**s). So we have to be careful not to select lines that are too close together.

In the case of the plane of parallel lines, we can specify a line at infinity. If that line is an **R**, we know it is equivalent to a translational degree of freedom.

In the next chapter, we encounter the situation of a C located at infinity.

In the general case of two lines in space (either Cs or Rs) we have the possibility for two lines to be skew, in which case their equivalent pattern is a "cylindroid." Helical degrees of freedom are introduced. They are the residents of the cylindroidal surface, which is defined by two skew Rs.

Compound connections (between more than two bodies) were explored along with a method for simplifying the complicated constraint patterns that occur between several component parts in a compound connection. We discovered that the mechanical connection between two bodies can be direct (parallel) or cascaded (series) through one or more intermediate bodies.

The net connection is obtained by adding the Cs if direct; adding the Rs if cascaded.

At the end of Chapter 3, we can look back on what has been learned so far and observe a profound symmetry between constraints and freedoms. Rules about patterns of C lines turn out to be applicable to patterns of R lines and vice-versa. This was not at all obvious at the outset. This will provide us with insights into machine design that are not available to someone using normal intuition.

Most of the basic principles of Exact Constraint Design have now been presented. These principles enable us to analyze mechanical connections using line patterns representing Cs and Rs. This analysis technique is called constraint pattern analysis. The technique is based on transformation of line patterns of two basic types: Rs and Cs. The transformations are governed by simple rules that permit manipulation of a pattern of one type (R or C) of line to an equivalent pattern of the same type or to a complementary pattern of the opposite type. The rules that govern these transformations are elegantly simple, and beautifully symmetrical. In fact, it is the symmetry that I find most remarkable. As we progress through the material of the ensuing chapters, we will discover still more symmetrical aspects between R patterns and C patterns. I hope you find this as fascinating as I do.

CHAPTER 4

Flexures

The flexure is a constraint device. We saw flexures in section 2.6 among the other constraint devices listed there. But because flexures seem to suffer more than other constraint devices from misuse and misunderstanding by designers, they need to be covered in a more thorough manner. Also, we will find that the techniques we use for analyzing flexure connections will be the basis for our study of structures in Chapter 7. For these reasons, the subject of flexures is now developed more thoroughly.

By using flexures as mechanical connections in the design of machines, a designer can often achieve a lower cost, higher performance solution than he or she would be able to achieve with any other type of connection. Among the flexure's attributes are zero friction, zero play, and low cost. The flexure's most obvious limitation is that it allows only small motions. Another "drawback" of flexures is that they need to be more carefully "engineered" than other constraint devices. They need to be thin enough to provide freedom of motion yet not so thin that they can be damaged by incidental worst-case loads. Many designers consider using flexures in machines that must make small, precise motions. But these machines are not the only candidates for the use of flexures. There are many instances where a body must be constrained by a particular connection in certain degrees of freedom. Often, a flexure can conveniently provide just those needed constraints.

4.1 ■ IDEAL SHEET FLEXURES

It is unfortunate that the term "flexure spring" occurs so frequently in the literature. A spring is exactly what an ideal flexure is *not*. Whereas "spring" suggests a device having some moderate level of stiffness, being somewhere between "rigid" and "compliant," a flexure *intends* to be as stiff as possible in certain directions while being as flexible as possible in other directions. Whenever we design with flexures, it is helpful to think of them as *ideal* flexures, exhibiting *binary* stiffness characteristics. Think of them as *absolutely rigid* in certain directions and *absolutely compliant* in other directions.

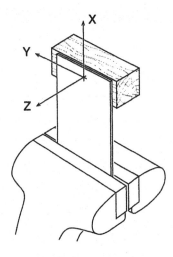

FIGURE 4.1.1

Slender bars (called wires) and thin plates (called sheets), although very compliant in flexure, are very stiff in tension and compression. The ratio of bending stiffness to stretching stiffness for these elements can be orders of magnitude.

To illustrate this, let us do an experiment. Glue a small block of wood to the edge of a 3 × 5 card and clamp the opposite edge of the card in a vise, as shown in Fig. 4.1.1. Now, let us see which of the block's six kinematic degrees of freedom have been constrained by our 3 × 5 card sheet. Observe that the block can be easily deflected back and forth along the Z-axis. Now, using the same amount of force, try to deflect the block along the X-axis or along the Y-axis. No observable deflection occurs. If we measure the stiffness along the Z-axis and compare it to the stiffness along the X or Y axes, indeed we would find that they are different by several orders of magnitude. This enormous ratio suggests that we can ignore the stiffness of our 3 × 5 card sheet along the Z-axis compared with the stiffness along the X and Y axes. To put it another way, we can say that our sheet imposes constraint along the X and Y axes, but does not impose constraint along the Z-axis. The block of wood is free to move along the Z-axis for small deflections. It is easy to see why this is so. X- and Y-axis forces, lying within the plane of the sheet, are trying to *stretch* or *compress* the sheet, whereas Z-axis forces are trying to *bend* it. Because the sheet is so thin, it bends very easily.

We can do a quick calculation to compare the stiffness of our sheet flexure in the X and Z directions.

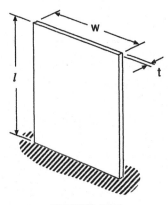

FIGURE 4.1.2

The stiffness in X:

$$k_x = \frac{AE}{l} = \frac{wtE}{l}$$

The stiffness in Z:

$$k_z = \frac{3\left(\frac{wt^3}{12}\right)E}{l^3} = \frac{wt^3E}{4l^3}$$

The ratio of

$$\frac{k_x}{k_z} = \frac{\frac{t}{l}}{\frac{1}{4}\left(\frac{t}{l}\right)^3} = 4\left(\frac{l}{t}\right)^2$$

For our 3×5 card, where

$$\frac{l}{t} \approx 300, \frac{k_x}{k_z} \approx 360,000$$

That is over five orders of magnitude!

The stiffness ratio does not depend on the modulus of elasticity, E, of the material, so it does not matter if our sheet flexure is made of cardboard or steel.

Now let us examine the block's rotational degrees of freedom. By doing a similar experiment, we arrive at the conclusion that the sheet constrains the block against rotation about the Z-axis, but does not constrain rotation about the X or Y axes.

We can summarize our observations about our sheet flexure model in the table of Fig. 4.1.3. Here we conclude that our sheet flexure is relatively stiff in X, Y, and θ_z, where the applied forces lie in the plane of the sheet. On the other hand, we note that when the applied forces lie outside the plane of the sheet, as with deflections in Z, θ_x, and θ_y, the sheet flexure is relatively compliant.

In summary we can say that our sheet flexure approximates an *ideal sheet flexure*, which we now define:

	Stiff	Compliant
X	✔	
Y	✔	
Z		✔
θx		✔
θy		✔
θz	✔	

FIGURE 4.1.3

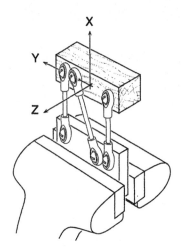

FIGURE 4.1.4

An *ideal sheet flexure* imposes absolutely rigid constraint in its own plane (X, Y, and θ_z), but it allows three degrees of freedom: Z, θ_x, and θ_y.

Figure 4.1.4 illustrates the bar equivalent of the ideal sheet flexure. It imposes exactly the same degrees of constraint and allows exactly the same degrees of freedom. The bars are connected at their ends with "ideal" ball joints. These ball joints are assumed to be frictionless and free of play. They cannot transmit any torque. Clearly, this is the "bar equivalent" of our sheet flexure.

Because both the sheet and bar structures exhibit the same kinematic behavior, they are functionally interchangeable. One is the functional equivalent of the other.

Figure 4.1.5 shows the constraint pattern of the sheet flexure, three nonredundant coplanar **C**s.

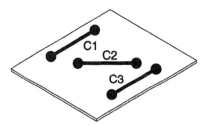

FIGURE 4.1.5

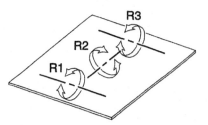

FIGURE 4.1.6

Indeed, we can imagine these three **C**s to lie along each of the bars in Fig. 4.1.4.

The complementary pattern of three coplanar **C**s is three coplanar **R**s in the same plane. Thus, a sheet flexure connection provides exactly the degrees of freedom shown in Fig. 4.1.6. This, of course, is no surprise. We found just these degrees of freedom in testing our sheet flexure model.

The freedom pattern of Fig. 4.1.6 might bring to mind the mechanism of Fig. 4.1.7. This mechanism consists of three coplanar hinges connected in cascade. This connection is also equivalent to the ideal sheet flexure, and to the three-bar connection of Fig. 4.1.4.

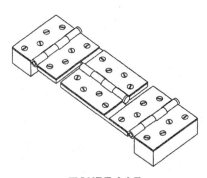

FIGURE 4.1.7

4.2 ■ CARDBOARD MODELING

The use of a cardboard sheet model is more than just a mental imaging technique for introducing the "binary constraint" nature of sheet flexures. It is a practical design tool that can be used by the designer in exploring new design ideas. The nice thing about the cardboard flexures is that they are simple, cheap, and fast to make. It is an incredibly easy, yet powerful technique for trying out various design configurations right in your office using a hot-melt gun and a pair of scissors for tools. And then, when you are satisfied with the shape and configuration that you have designed to fit your application, you do a few engineering calculations and the design is done. Then you can build it in metal.

4.3 ■ IDEAL WIRE FLEXURES

Now let us do another experiment, similar to the one we did with the sheet clamped in the vise, but this time we will use a slender bar or wire instead of a sheet. If we test the stiffness of the block's position in each degree of freedom, we discover that it is very free to move in every degree of freedom except X. If we try to pull the block up in X (loading the wire in tension), we find no observable deflection. The same is true for a downward force applied to the block (loading the wire in compression) as long as we do not load the wire so heavily that we cause it to buckle.

When the experiment is completed, we can observe that the wire flexure offers *orders of magnitude* higher stiffness along its axis compared with any other direction. We can thus conclude that a slender bar or wire approximates an "ideal wire flexure."

An *ideal wire flexure* imposes an absolutely rigid constraint along its axis (X), but it allows five degrees of freedom: Y, Z, θ_y, θ_z, θ_x.

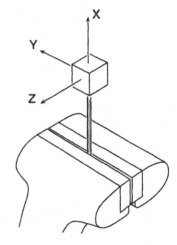

FIGURE 4.3.1

An ideal wire flexure is kinematically equivalent to a bar with a ball-and-socket joint at each end. This is the conceptual model a designer should have when designing with wire flexures.

4.4 ■ PRACTICAL TIPS ON FLEXURE CONNECTIONS

Use Clamp Plates When Attaching Sheet Flexures

A clamp plate eliminates the possibility that the flexure will become distorted by the rotary motion of the screw heads as the screws are tightened. It also ensures that a uniform, flat clamping effect is achieved. The likelihood of having stress concentrations at the screw holes is reduced. Applied forces result in uniform stresses distributed across the full width of the flexure.

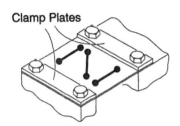

FIGURE 4.4.1

Always Use Flexures in Their Nominally Flat (Straight) Condition, Never Substantially Curved

This advice goes back to the experiments we did in section 4.1. The extremely high stiffness ratio that we observed between "in-plane" and "out-of-plane" forces owed to the flat condition of the sheet and to the straight condition of the wire. If the sheet or the wire had been substantially curved, then we would not have observed their "binary constraint" property.

Figure 4.4.2 shows a body constrained by a ball joint at one end and a sheet flexure at the other. Satisfy yourself that the constraint pattern provided by these two constraint devices is proper, and that the body is exactly constrained.

Observe that the ball, in order to nest in a conical hole, must have a vertical nesting force.

When designing a connection such as this, there is a great temptation to use the sheet flexure as a leaf spring to provide a nesting force for the ball-and-socket joint. *AVOID THE TEMPTATION.* To be an effective constraint device, a flexure must be nominally flat so that the equivalent bars are *straight*. A flat leaf spring, on the other hand, must be bent into a *curved* shape in order to generate a force. These two purposes are at odds. Therefore, when the utmost in precision is required, a flexure should not be required to perform double duty as a spring. A separate spring should be used.

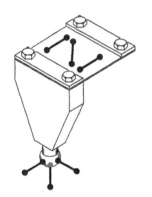

FIGURE 4.4.2

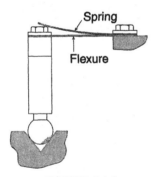

FIGURE 4.4.3

No Bend at Attachment

In the interest of achieving a simpler design with fewer parts, a designer might be tempted to combine two flexures into one, as shown in Fig. 4.4.4. But beware of the performance compromise that is made. A small (but for some purposes, significant) deflection will occur at the bends as the load changes. This has the effect of drastically reducing the stiffness of the connection, compared with the "straight flexure" configuration shown in Fig. 4.4.6. A second disadvantage of the bent flexure design is that fully hard material cannot be used because it might fracture at the bend.

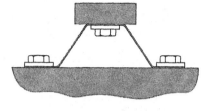

FIGURE 4.4.4

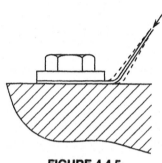

FIGURE 4.4.5

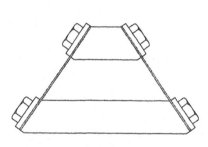

FIGURE 4.4.6

Attaching Wire Flexures

The biggest challenge with wire flexures is in achieving a simple, effective attachment. Here are a few different ways that I have come across.

Fig. 4.4.7 shows a simple loop in the end of the wire.

In Fig. 4.4.8, the wire nests in a shallow "V" groove and is then clamped in place. Sharp ridges in the underside of clamp plate "bite" into wire.

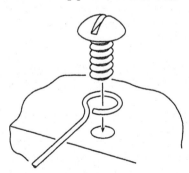

FIGURE 4.4.7

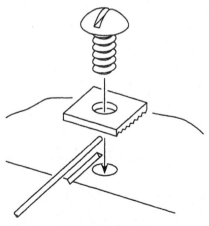

FIGURE 4.4.8

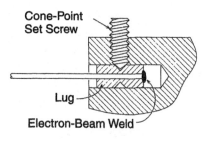

FIGURE 4.4.9

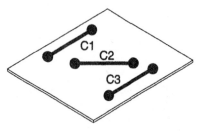

FIGURE 4.4.10

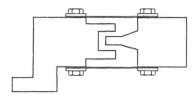

FIGURE 4.5.1

In Fig. 4.4.9, lugs have been soldered or (electron-beam) welded to the ends of the wire. Each lug is then secured in a hole with a cone-point set screw.

Provide Limit Stops to Prevent Accidental Overtravel

If a flexure design is to be successful, care must be taken to ensure that the level of stress developed in the flexure(s) does not exceed a level that will result in damage. Stress levels produced by repeated "normal" deflections must be kept below a stress value which has been determined based on the material properties, total number of cycles of operation, safety factor, and so on. In addition to so-called "normal" loads," the designer must be aware of the possibility that the flexures may sustain occasional "worst-case" loads as a result of shipping, mishandling, or machine malfunction. Clearly, a considerable amount of thorough stress analysis must be done to ensure that a flexure design is "robust." This analysis, however, is beyond the scope of Constraint Pattern Analysis.

It is often possible to design the shape of two bodies connected by flexures so that features on the two bodies will be clear of each other during normal operation but will make contact in the event of overtravel. Once the features have made contact, further deflection of the flexures is prevented and damage is avoided. These features are called limit stops. They may be fixed or adjustable.

4.5 ■ PATTERNS OF Cs AND Rs THAT OCCUR IN FLEXURE CONNECTIONS

The two basic flexure elements are the sheet and the wire. The constraint pattern for the wire is a single constraint along the wire.

> A single straight-wire flexure provides a single C along its axis.

The constraint pattern for the sheet is three constraints in the plane of the sheet. The three Cs can be located somewhat arbitrarily, so long as they do not approach the condition of overconstraint, where they all intersect one point.

> A single flat-sheet flexure provides three Cs in the plane of the sheet.

In the examples shown in Figs. 4.5.2 and 4.5.3, we have two sheet flexures connected *directly* between bodies A and B. This is analogous to the *parallel* connection in the electrical resistance network described in section 3.7. Be careful to understand that the two sheet flexures of Fig. 4.5.2 are said to be *connected in parallel* between bodies A and B in the sense that their connection is analogous to electrical circuit components connected in *parallel*. The flexures themselves are very clearly not parallel to each other in a geometric sense.

When two sheet flexures are connected directly (in parallel) between two bodies, as in Fig. 4.5.2, the Cs of each sheet are added. Each sheet contributes three Cs, for a total of six, but one is redundant. It is not possible to have six constraints in two planes without overconstraint. (See section 3.6.) Thus the connection has five Cs. The complementary pattern is a single R located at the intersection of the planes of the two sheets.

> Two flat sheet flexures connected in *parallel* (directly connected) define a single R located at the intersection of their two planes.

In the case where the two flexure planes are parallel, as in Fig. 4.5.3, the intersection occurs at infinity. Of course, the R at infinity is equivalent to a translation perpendicular to the flexure planes.

One of the symmetries that exists between Rs and Cs relates to sheet flexures. The sheet flexure, which is represented as three Cs lying in the plane of the flexure, can be equivalently represented as three Rs lying in the plane of the flexure. As with the Cs, the locations of these three Rs can be located somewhat arbitrarily, anywhere in the plane of the sheet without redundancy.

> A single flat-sheet flexure can be equivalently represented as three Rs in the plane of the sheet.

If two sheet flexures are cascaded (*connected in series*) as in Fig. 4.5.5, the Rs of each of the sheets are added. The result is a single C at the intersection of the planes of the two sheets. Again, we see symmetry with the case where two sheets are connected in parallel. The parallel connection of two sheets pro-

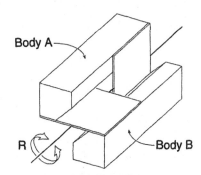

FIGURE 4.5.2

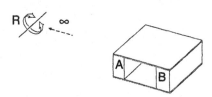

FIGURE 4.5.3

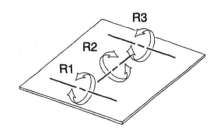

FIGURE 4.5.4

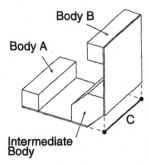

FIGURE 4.5.5

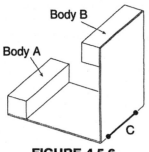

FIGURE 4.5.6

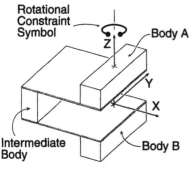

FIGURE 4.5.7

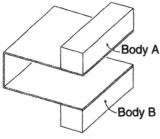

FIGURE 4.5.8

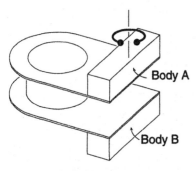

FIGURE 4.5.9

duces an $\mathbf{R}$ at their intersection. The series connection of two sheets produces a $\mathbf{C}$ at their intersection.

Two flat-sheet flexures *connected in series* define a single $\mathbf{C}$ located at the intersection of their two planes.

Consider the configuration of Fig. 4.5.6, where the intermediate body of the previous example has been replaced by a fold. It turns out that every point along the fold line has a fixed, rigid relation to every other point along the fold line. This is because the material of each adjoining flexure leaf prevents relative motion between these points in its respective plane. (This will be revisited in Chapter 7.) As a result, the fold itself behaves like a rigid intermediate body. Thus, the folded sheet flexure will provide a single $\mathbf{C}$ along the fold line.

Next is the case where two parallel sheets are connected in series. We know what to expect. We expect a $\mathbf{C}$ at the intersection of the two sheets. But the sheets intersect at infinity. This $\mathbf{C}$ at infinity is equivalent to a pure rotational constraint perpendicular to the planes of the sheets. Just as we earlier found that an $\mathbf{R}$ at infinity is equivalent to a pure translation $\mathbf{T}$, we now discover that a $\mathbf{C}$ at infinity is equivalent to a pure rotational constraint. We use the symbol shown in Fig. 4.5.7 to represent rotational constraint.

Care must be taken to ensure that the intermediate body of Fig. 4.5.7 is *rigid*, not flexible. For example, a simple sheet bracket with two bends, like the one shown in Fig. 4.5.8, would not work. The analysis of this configuration reveals *three* sheet flexures connected in cascade. This provides a pattern of three $\mathbf{R}$s in each of three planes. There is no line in space that can be found to intersect each of these nine lines, therefore, the connection provides zero constraint.

By contrast, however, consider the configuration shown in Fig. 4.5.9, which consists of two parallel sheets joined by a thin-walled tube. In Chapter 7, we will learn that the tube can be a three-dimensionally rigid shape. Because the two flexure sheets are connected by a rigid intermediate body, the connection shown in Fig. 4.5.9 is equivalent to that shown in Fig. 4.5.7.

4.6 ■ DEFLECTED SHAPE OF A SHEET FLEXURE

As we have already discussed, whenever a sheet flexure is being used, it is advisable to use it in its flat condition to take full advantage of the high in-plane stiffness it provides. However, there are many instances (such as in mechanisms) where flexures are required to undergo some deflection. For sheet flexures, this deflection occurs in any of *three modes*. Each of these modes corresponds to one of the sheet flexure's three degrees of freedom. The first mode is a simple bend. This is the deflection mode of the sheet flexures of the well-known "crossed-flexure" hinge connection shown in Fig. 4.6.1. The two connected bodies are "hinged" about a line defined by the intersection of the two flexure planes. Typically, an angular excursion of 10 or 20 degrees is easily achieved between the two bodies.

The deflected shape of the flexures in Fig. 4.6.2 is an "S" bend. This is the second deflection mode. If the flexures in this device are limited to the same maximum stress as those of Fig. 4.6.1, then the bodies in Fig. 4.6.2 would be capable of only a much smaller angular excursion.

The third deflection mode of a sheet flexure is twist. Fig. 4.6.3 shows a sheet flexure subjected to twist. The angular excursion permitted between these two bodies might be only a couple of degrees.

To increase the allowable excursion of the bodies connected by flexures, a designer might alter the dimensions of the flexures by making them longer or thinner. But this will make the flexures more vulnerable to elastic buckling. On the other hand, in a mechanism where only small excursions are needed, we might consider "beefing up" the flexures by making them shorter and thicker.

The "monolithic" flexure of Figure 4.6.4 represents the *limit* in making a sheet flexure short and thick. It will not elastically buckle. Also, it will have only one mode of deflection: simple bending. This deflection will have a very limited range, perhaps only a fraction of a degree, depending on construction material.

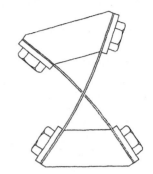

FIGURE 4.6.1

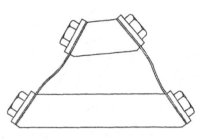

FIGURE 4.6.2

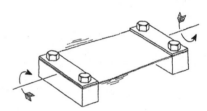

FIGURE 4.6.3

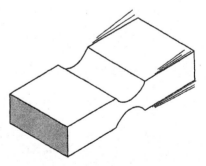

FIGURE 4.6.4

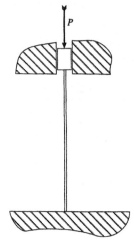

FIGURE 4.7.1

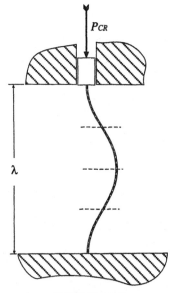

FIGURE 4.7.2

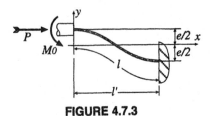

FIGURE 4.7.3

4.7 ■ ELASTIC BUCKLING

Fig. 4.7.1 shows an edge view of a flat sheet flexure or a straight-wire flexure subjected to compression load, P. The flexure will rigidly bear the load P until the magnitude of P reaches the value P_{cr}, the threshold of elastic buckling.

Once P_{cr} has been exceeded, the flexure is *buckled*. The buckled flexure then takes on the shape shown in Fig. 4.7.2. This shape is one period of a cosine wave of wavelength λ. As the load is further increased, the amplitude of the cosine wave becomes greater, but the stiffness of the flexure is now drastically reduced. In order for the flexure to provide high stiffness, we must ensure that the applied load does not exceed P_{cr}.

Most engineering textbooks present the equations that describe elastic buckling but do not describe the phenomenon in terms of the deflected shape of the flexure. I think this is unfortunate because it leaves engineers with only a cookbook approach to solving buckling problems. By understanding the phenomenon in terms of the flexure's shape, a more thorough understanding is achieved.

Let us now analyze the deflected shape of the buckled flexure in response to the application of compression load P at values *above* P_{cr}. We will examine the flexure shown in Fig. 4.7.3, whose length is just one half that of the flexure shown in Fig. 4.7.2. M_0 is the moment that accompanies load P, maintaining zero slope at the left end.

l = actual length of flexure
l' = projected length parallel to load P
e = deflection, perpendicular to load P

The moment at any point x along the flexure is

$$M = -M_0 + P\left(\frac{e}{2} - y\right)$$

From beam deflection theory, we know that

$$\frac{d^2y}{dx^2} = \frac{M}{EI}$$

$$\frac{d^2y}{dx^2} = \frac{M}{EI} = \frac{1}{EI}\left[-M_0 + P\left(\frac{e}{2} - y\right)\right] \qquad (1)$$

This requires that

$$y = \frac{e}{2} \cos\left(\frac{\pi x}{l'}\right) \qquad (2)$$

$$\frac{dy}{dx} = -\frac{e\pi}{2l'} \sin\left(\frac{\pi x}{l'}\right) \qquad (3)$$

$$\frac{d^2y}{dx^2} = -\frac{e}{2}\left(\frac{\pi}{l'}\right)^2 \cos\left(\frac{\pi x}{l'}\right)$$

$$\frac{d^2y}{dx^2} = -\left(\frac{\pi}{l'}\right)^2 y$$

Substituting this into Eq. (1) gives

$$-\left(\frac{\pi}{l'}\right)^2 y = \frac{1}{EI}\left[-M_0 + P\left(\frac{e}{2} - y\right)\right]$$

$$y\left[\left(\frac{\pi}{l'}\right)^2 - \frac{P}{EI}\right] = \frac{1}{EI}\left(M_0 - P\frac{e}{2}\right)$$

Because y is a cosine function and the right-hand side is not, the only way the equation can be true is if both sides equal zero.

$$\left(\frac{\pi}{l'}\right)^2 - \frac{P}{EI} = 0 \qquad\qquad M_0 - P\frac{e}{2} = 0$$

$$l' = \pi\sqrt{\frac{EI}{P}} \quad (4) \qquad\qquad M_0 = P\frac{e}{2} \quad (5)$$

Several interesting things have now been discovered about the compression-loaded flexure in the region $P > P_{cr}$:

A. The shape of the flexure follows a half wave of a cosine curve having a wavelength 2 l' and amplitude $e/2$ [Eq. (2)].

B. When the flexure is straight ($e = 0$) $l' = l$ and Eq. (4) is exactly the Euler formula for column buckling.

C. The flexure can support loads in *excess* of P_{cr}. Loads greater than P_{cr} are elastically supported by the flexure, whose shape is ½ period of a cosine wave. As the load is increased, the wavelength of the cosine curve, 2 l', is reduced. The amplitude of the cosine curve, $e/2$, increases.

D. The bending moment is zero at the mid-point of the curve (flexure) and maximum at the ends [Eqs. (1) and (5)].

Because the moment at the mid-point of the flexure equals zero, an *equivalent loading* could be drawn, as in Fig. 4.7.4. This is a ¼ wave of the basic cosine curve and leads to an interesting experiment.

Imagine a lead ball of weight P supported at the end of a slender pole that is sticking straight up out of the floor, as shown in Fig. 4.7.5. Now imagine that a mechanism beneath the floor is advancing the pole upward so that the length of the pole, h, between the weight and the floor is growing.

We observe that for a large range of length of the pole, the height of the weight above the floor remains constant. The height is

$$h = \frac{l'}{2} = \frac{\pi}{2}\sqrt{\frac{EI}{P}}$$

The height depends only on

• the modulus of elasticity of the pole material
• the moment of inertia of the pole
• the size of the weight at the top

If the length of the pole is less than $\frac{\pi}{2}\sqrt{\frac{EI}{P}}$, the pole will remain perfectly straight. But once the length of the pole exceeds $\frac{\pi}{2}\sqrt{\frac{EI}{P}}$, the pole is "critically loaded" and takes on the shape of a quarter wave of the cosine curve.

As the pole is advanced still further out of the floor, the height h of the weight remains constant and

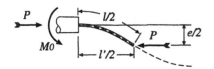

FIGURE 4.7.4

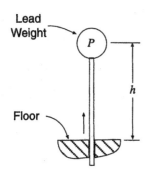

FIGURE 4.7.5

the amplitude of the cosine wave grows. The weight simply moves laterally at a constant height.

Eventually, the stresses in the extreme fibers of the pole will exceed the elastic limit of the material. When the pole fails, the weight will finally drop.

This experiment is interesting because it demonstrates the elastic nature of Euler column buckling. It is not a catastrophic failure. By using limit stops, it is possible to endure buckling due to occasional overloads without ruining the flexure. We also see from this experiment that by changing the end constraint conditions of the flexure, the deflected shape of the flexure follows a different portion of the cosine curve. By knowing the end constraint conditions on any given flexure, a designer can relate the flexure length to the length l of the Euler equation and correctly find the value of P_{cr}. Here are some additional end-constraint conditions with the corresponding buckled shape. In each case, the buckled shape is some portion of the cosine curve shown in Fig. 4.7.2.

For example, Fig. 4.7.6 shows a buckled flexure whose ends are both constrained laterally but are free rotationally. The deflected shape of this flexure will follow the center section of the cosine curve.

The flexure shown in Fig. 4.7.7 will also follow one half of the cosine curve, but it will be a different

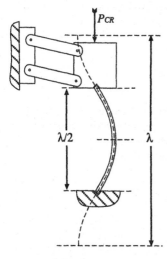

FIGURE 4.7.6

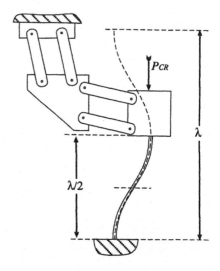

FIGURE 4.7.7

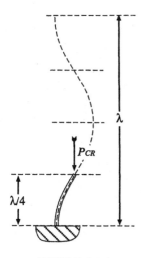

FIGURE 4.7.8

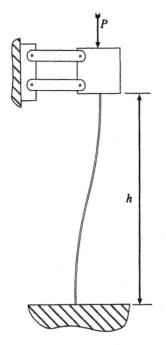

FIGURE 4.7.10

portion of the curve than in the previous case. In this case, both ends of the flexure have been constrained laterally but are free rotationally.

Figure 4.7.8 shows a flexure in which one end is free both laterally and rotationally. The deflected shape of this flexure follows one quarter of the cosine curve. (This is the loading condition shown in Fig. 4.7.4.)

Compare the shape of the buckled flexure shown in Fig. 4.7.8 with that of the deflected cantilever beam of Fig. 4.7.9. Although both deflected shapes *look* the same, they are subtly different. The cantilever beam follows a polynomial curve as a result of

FIGURE 4.7.9

load F.

Now examine the deflected flexure shown in Fig. 4.7.10. When $P = 0$, this flexure has a shape that is the same as two deflected cantilever beams joined at their free ends. When load P is increased, the shape of the flexure undergoes a subtle change, but only a very small vertical deflection will occur at the upper end. When the value of load P reaches

$$P_{cr} = \frac{4\pi^2 EI}{h^2}$$

the deflected flexure will then follow a shape that is *one period* of a cosine curve. Does it *look* like the deflected flexure of Fig. 4.7.10 follows one period of a cosine curve?

No, it is not obvious by looking at the deflected flexure what its shape is. In fact, it would be dangerous to presume to know its shape just by looking at it because it *looks* like *one half period* of a cosine wave. But the *constraint condition* imposed on the flexure dictates that the flexure will follow *one period* of a cosine wave, as shown in Fig. 4.7.11.

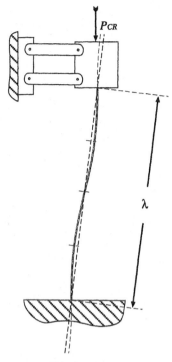

FIGURE 4.7.11

CHAPTER SUMMARY

By using the Constraint Pattern Analysis technique, we have analyzed sheet and wire flexure connections between bodies. Analysis of familiar connections, such as the "crossed-strip" flexure configuration, gives us exactly the results we expected. Next we used the techniques to analyze other, unfamiliar configurations for which our intuition gave us either no answer or the wrong answer. Constraint Pattern Analysis gives us the net pattern of constraints or freedoms in a remarkably fast, easy solution.

CHAPTER 5

Couplings

Shaft couplings are used to provide rotational constraint connection between two shafts that are approximately coaxial. There are many different shaft couplings on the market from which a designer may choose. So many, in fact, that selection of the "best" one for a given application can be confusing.

However, if the designer first diagrams the pattern of Cs which have already been imposed on the two shafts, he or she will then be in a position to know just what pattern of Cs (or Rs) is needed from the coupling.

Once the field of prospective couplings has been narrowed to only those providing the correct pattern of Cs (Rs), the selection can then be further refined on the basis of an application's specific requirements, such as cost, backlash, maximum allowable misalignment, windup stiffness, strength, maintenance, and so on.

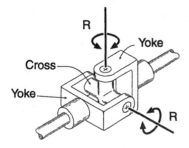

FIGURE 5.1.1

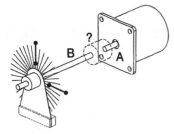

FIGURE 5.1.2

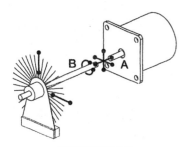

FIGURE 5.1.3

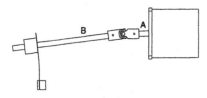

FIGURE 5.1.4

5.1 ■ FOUR-C COUPLINGS

The universal joint coupling (U-joint) shown in Fig. 5.1.1 is a well-known shaft coupling. It consists of three major parts: two yokes and a "cross." The parts are connected in series (cascaded), resulting in two crossed Rs, $R1$ and $R2$. The complementary pattern is four Cs, as shown in the diagram adjacent to the title of this section. This pattern of Cs suggests this coupling will provide a "ball-and-socket"-like connection between the two shafts *plus* a rotational constraint about the shaft axes.

The rotational constraint gives the coupling a characteristic that is known as high *windup* stiffness. Generally, this is a desirable characteristic of a shaft coupling.

Consider the situation shown in Fig. 5.1.2, where shaft A is the shaft of a motor, whose position is fixed. Shaft B is supported at its left end in a flexure mounted plain bearing. The flexure mounted plain bearing provides shaft B with two radial Cs lying in the plane of the flexure and intersecting at the shaft axis.

A four-C coupling (such as a U-joint) is just what is needed to achieve an exact constraint connection to shaft B. The four-C pattern of the coupling, supplemented by the two-C pattern from the flexure mounted bearing exactly constrains each of shaft B's six degrees of freedom. Fig. 5.1.3 shows the constraint diagram for this connection.

By designing such an exactly constrained connection to shaft B, the apparatus is insensitive to inaccuracies of position between the motor and the bearing supporting shaft B. Fig. 5.1.4 shows an exaggerated view of the apparatus in which the bearing is mispositioned. Shaft B ends up being angularly misaligned with shaft A. Rs in the coupling and in the flexure/bearing connection easily accomodate this misalignment.

5.2 ■ THREE-C COUPLINGS

The three-C coupling provides a pattern of three coplanar Cs, where the plane of the Cs is perpendicular to the axes of the connected shafts. You may recall that this is precisely the constraint pattern imposed by a sheet flexure. The complementary pat-

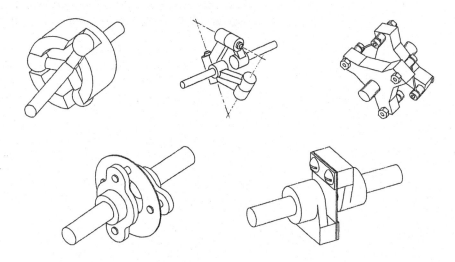

FIGURE 5.2.1

tern is three **R**s in the same plane. We can think of many coupling configurations that give us this pattern.

A three-**C** coupling would be specified to connect the shafts of the apparatus of Fig. 5.2.2, where shaft A is fixed (mounted in two bearings) and shaft B is mounted in only one bearing. Here, the possibility of angular misalignment must be accommodated.

The constraint pattern analysis shows that shaft B receives three **C**s from its single bearing and three **C**s from the coupling connection to shaft A.

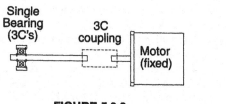

FIGURE 5.2.2

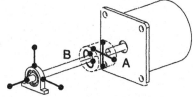

FIGURE 5.2.3

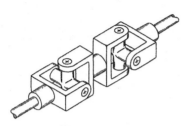

FIGURE 5.3.1

5.3 ■ TWO-C COUPLINGS

Consider the connection sketched in Fig. 5.3.1, which consists of two cascaded universal joints.

Clearly, each U-joint is, itself, a cascaded connection, so the entire connection provides four **R**s between the two shafts. These degrees of freedom are shown in Fig. 5.3.2.

Note that this pattern of four **R**s consists of two intersecting pairs of lines. The complementary pat-

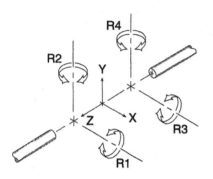

FIGURE 5.3.2

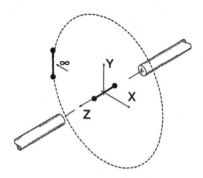

FIGURE 5.3.3

tern is easily found to be two **C**s: one **C** joining the two intersection points (along the shaft axis), and another **C** at the intersection of the two (parallel) planes. Because the plane of pair **R**1 and **R**2 is parallel to the plane of pair **R**3 and **R**4, the intersection occurs at infinity. This **C** at infinity is positioned tangentially on a circle of infinite radius. The circle lies in the X–Y plane with its center at 0,0. This **C** at infinity is a θ_z rotational constraint. It prevents rotation about the Z axis.

Alternatively, this constraint pattern can be expressed as in Fig. 5.3.4.

Now recall Fig. 5.3.2. Notice that **R**1 and **R**3 are parallel and therefore define a horizontal plane of equivalent parallel lines. **R**2 and **R**4 are also parallel, defining a second plane of parallel lines. We are free to choose any two lines from each plane and use them to represent the degrees of freedom provided by the coupling of Fig. 5.3.1. For example, we would be quite correct to say that the four degrees of freedom of Fig. 5.3.1 are those shown in Fig. 5.3.5.

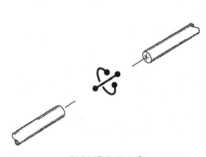

FIGURE 5.3.4

Thus, Fig. 5.3.5 is an equivalent representation of the degrees of freedom provided by the double U-joint coupling, where each of the **T**s is equivalent to an **R** at infinity. When the freedoms of the coupling are expressed this way, it brings to mind the universal, lateral coupling of Fig. 5.3.6. The universal-lateral coupling is a cascaded connection from hub to ring to hub where two degrees of freedom (sliding and rotation about each transverse axis) are permitted in each connection.

This simple constraint pattern analysis has shown us that for small motions, the universal-lateral coupling is kinematically equivalent to the double U-joint.

Connection between two misaligned shafts is the most obvious use of a coupling, but not the only important use. We will now see that a two-**C** coupling is exactly what is needed to provide the proper nut-to-carriage connection in a lead screw driven carriage.

Figure 5.3.7 is a cutaway view detailing the construction of a fairly typical lead screw driven translation stage. In this device, the carriage is supported on a pair of rails, providing constraint in all but one degree of freedom. The carriage remains free in X. (Actually, the rails provide *overconstraint* in three degrees of freedom, but let us overlook that now to focus on the lead screw and nut.) Because the lead screw is mounted to the base by a pair of ball bearings, its position relative to the base is fixed in X, Y, Z, θ_y, and θ_z. Furthermore, some drive means, such as a motor (not shown), provides θ_x constraint to the lead screw. Thus when the nut, which is threaded onto the lead screw, is also rigidly attached to the carriage, it is clear that a condition of gross overconstraint has been achieved. The position of the nut in Y, Z, θ_y, and θ_z has been constrained by *both* the lead screw *and* the carriage.

The *proper* connection between the nut and carriage, however, must provide only two constraints. This conclusion should not be too surprising considering that the job of the lead screw and nut is to provide an adjustable X constraint between the base and carriage, nothing more, nothing less. The position of the nut in X is linearly related to the θ_x position of the lead screw, provided the nut's θ_x position and the lead screw's X position are both fixed. Thus, we conclude that in addition to the X constraint needed between nut and carriage, a θ_x constraint is also needed. No other constraints are needed or wanted. The connec-

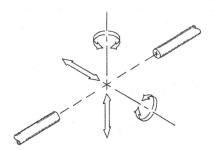

FIGURE 5.3.5

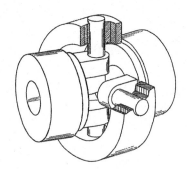

FIGURE 5.3.6

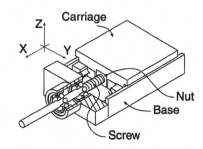

FIGURE 5.3.7

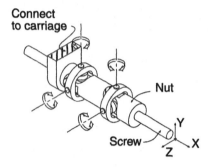

Connect
to carriage

Nut

Screw

FIGURE 5.3.8

tion between nut and carriage should be free in Y, Z, θ_y, and θ_z because the lead screw already constrains the nut in these four degrees of freedom.

There is no pattern in which we can arrange just two constraints to provide constraint in X and θ_x. We must resort to some sort of cascaded connection. For example, we can cascade two connections each having two degrees of freedom or four connections each having a single degree of freedom to achieve the necessary four degrees of freedom. One such connection, familiar to us as a shaft coupling, is illustrated in Fig. 5.3.8. This device contains two universal joints. (Ordinarily, each universal joint would consist of a cross and trunnion, but here rings have been used in place of the crosses because the lead screw goes through the center.) The pair of U-joints provide four rotational degrees of freedom. Thus, this connection provides exactly the two constraints needed.

If this is exactly the connection needed between the nut and carriage of the translation stage, why have we not seen it before? I cannot help wondering why the use of shaft couplings has become so widely accepted as good mechanical design practice for connecting two shafts but has only rarely been applied to the connection between a nut and carriage. I suppose it is just one of life's mysteries. The requirements for the two connections are identical. The nut-to-carriage connection requires two constraints, θ_x and X. This is exactly the connection imposed by a four-**C** shaft coupling.

The similarity between the shaft coupling and the nut-to-carriage connection might lead a designer to consider any number of different well-known shaft couplings to use as a nut-to-carriage connection. Figure 5.3.9 illustrates the use of a double-disk coupling for this purpose, but because this coupling provides only θ_x constraint, a separate X connection must also be added. Also of interest is that only a half-nut is used. An obliquely oriented force, F, provides both the nesting force for the half-nut against the lead screw as well as the nesting force for the X constraint. The advantage of using just a half-nut nested against the lead screw is in the precision of fit that can be

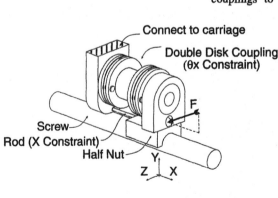

Connect to carriage

Double Disk Coupling
(θx Constraint)

Screw
Rod (X Constraint)
Half Nut

FIGURE 5.3.9

achieved between the nut and screw. As wear occurs, the thread form of the nut becomes more nearly matched to that of the screw.

The nut-to-carriage connection of Fig. 5.3.10 comprises two cascaded connections, each having two degrees of freedom. This device was designed by Brad Jadrich of Eastman Kodak Company for use in a sensitive scan printing apparatus. The flexure wires were specified to ensure that there would be no looseness or play in the connection (as there might be with a double U-joint connection).

Kinematically, all of these nut-to-carriage connections are equivalent. They each provide two constraints: X and θ_x. The trick is to devise a connection that both provides the necessary degrees of constraint and is easily and cheaply constructed. In that sense, perhaps the most elegant of these connections is that shown in Fig. 5.3.11. This device, disclosed in US Patent #3,831,460 (8/27/74), consists simply of a slotted tube, where the effect of the slots is to produce four monolithic flexure hinges (**R**s). The tube wall thickness is presumed to be great enough so that each of the sections of the tube (between slots) can be safely considered to be a separate rigid body.

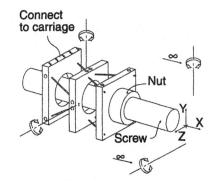

FIGURE 5.3.10

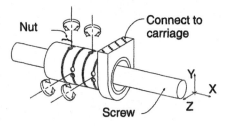

FIGURE 5.3.11

5.4 ■ ONE-C COUPLINGS

The configuration of Fig. 5.4.1 (repeated from Fig. 4.5.9) is the basic element of the bellows, a flexural configuration that is well known for providing pure rotational constraint. The bellows is actually a cascade of flat parallel sheets connected by rigid circular tubes. Although each of the flat plates in the bellows can accommodate only a small deflection, the entire structure can accommodate a fairly large deflection. However, it should be remembered that because the bellows consists of a *cascaded* (series) connection of sheet flexures, it is the three **R**s of each sheet (the five **R**s of each pair of adjacent sheets) that must be added. If the bellows is to maintain its high stiffness rotational constraint property without degradation, the planes of all of the sheets must intersect at a common line.

In other words, the axis of the bellows must follow a planar arc of constant radius of curvature. This can be easily understood in the following way. We

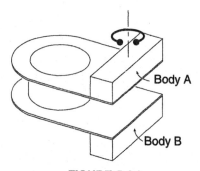

FIGURE 5.4.1

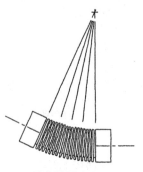

FIGURE 5.4.2

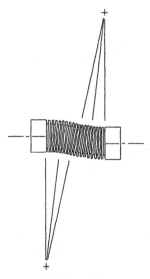

FIGURE 5.4.3

have already learned that each pair of adjacent sheet flexures defines a single constraint at the intersection of their planes. Conceptually, we could think of this constraint as being provided by a wire flexure. We could imagine this for each adjacent pair of sheet flexures in the bellows. Now, if we have lots of wire flexures connected together in series, the only way we can have a constraint through them all is if they are all collinear.

This is pretty important to know if we are planning to use the bellows to provide a rigid torsional coupling between two misaligned shafts. For example, the configuration shown in Fig. 5.4.3 will prove to be torsionally compliant, because the axis of the bellows does not follow a single circular path, but instead follows an "S"-shaped path.

This turns out to be a surprise to most mechanical designers. What is really amazing is that even the manufacturers of the bellows couplings are not aware of this. Their couplings are sketched in their catalogs just like Fig. 5.4.3 without any mention of the resulting penalty in windup flexibility.

I recently worked on a project where a flexible shaft was being considered as a coupling between a servomotor and a lead screw. A flexible shaft is like a speedometer cable without a sheath. Despite the lack of resemblance structurally to the bellows, it behaves kinematically just like a bellows. It is very stiff in windup as long as it follows a circular path. But if you bend it into an S-shaped path, it loses this stiffness. At first, a couple of the engineers on the project were reluctant about using the flexible shaft, because they remembered some bad experiences they once had with them. They, like most designers, were not aware of the important relationship between the shape of the flexed path and the torsional stiffness. After seeing a demonstration, they were willing to reconsider. This is a good example of the kind of insight that is available to mechanical designers when they use Constraint Pattern Analysis.

5.5 ■ ZERO-C COUPLINGS

Some couplings are made of rubber and provide no rigid constraint. Such a coupling, shown in Fig. 5.5.1, is flexible in every degree of freedom. Not only will this coupling accommodate shafts that are misaligned angularly, laterally, and axially, it will also permit rotational windup. In some servo applications, this would not be tolerable, but for other applications it may be desirable.

A coupling does not need to be made of rubber to be compliant in windup. Imagine coupling two shafts with a coil spring. Clearly, such a coupling will provide no rigid constraints. The coupling shown in Fig. 5.5.2 is really just a coil spring whose *coils* have been *cut* instead of wound. It would be a poor choice for an application requiring rigid constraint in windup.

FIGURE 5.5.1

FIGURE 5.5.2

CHAPTER SUMMARY

Shaft couplings are not all created equal. In this chapter, we learned that they come in five distinctly different categorical types, from four Cs to zero Cs. For mechanical designers, it can be confusing, at best, to sort among the myriad available couplings to select the "best" one for a particular application. By using Constraint Pattern Analysis techniques, we can separate the facts from the advertising hype and make a prudent choice. This technique can even lead to the invention of a new coupling design!

CHAPTER *6*

R/C Patterns in Hardware

In this chapter we explore the patterns of **R***s and* **C***s that exist in various mechanical connections. By looking beyond the hardware (constraint devices such as contact points, links, bearings, ball-and-socket joints, flexures, etc.) and examining the patterns of lines that represent the applied constraints and the resulting freedoms, we will discover a realm in which mechanical connections can be easily represented and visualized. By becoming facile with constraint diagramming techniques, the machine designer acquires a powerful tool for synthesizing and analyzing the design of a machine.*

6.1 ■ DEMOUNTABLE CONNECTIONS THAT PROVIDE RIGID, PRECISE NESTING (SIX Cs / ZERO Rs)

In this section we examine various configurations of six constraints exactly constraining every degree of freedom of a rigid body with respect to a reference body. In particular, we are interested in connections that are easily disassembled and reassembled and which, when reassembled, will return to an exactly repeatable (precise) position. Such connections generally consist of some arrangement of six contact points with a nesting force positioned to provide a normal force at each contact point. We refer to such a connection as a "demountable" connection.

As a first example of such an exactly constrained, demountable connection, consider the connection shown in Fig. 6.1.1. Here is a body exactly constrained by a pattern of six constraints: three parallel constraints against the bottom surface, two parallel constraints against a second surface perpendicular to the first, and a single constraint against a third surface, perpendicular to the other two surfaces. Each of the constraints is provided by a contact point. Using vector addition, a single nesting force vector has been found which will produce an equal normal force at each of the six contact points. This pattern of constraints, called the "3-plane" pattern, is sometimes used for holding parts precisely for machining or gaging. Dimensions to various features are then measured from three mutually orthogonal datum planes: the primary datum plane defined by the contact points of the three parallel constraints, the secondary datum plane defined to be perpendicular to the primary plane and containing the contact points of the two parallel constraints, and the tertiary datum plane defined to be perpendicular to the primary and secondary planes and containing the contact point of the last remaining constraint.

A practical example of the use of the 3-plane pattern of constraints is in the mounting of the solid glass beam splitter prism shown in Fig. 6.1.2. This prism can be easily removed for cleaning and then replaced into its exact initial position without any adjustment or tweaking. Two symmetrically placed springs combine to provide a nesting force that loads the prism against the three contact points of the primary plane and the two contact points of the secondary plane, but not

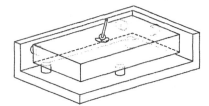

FIGURE 6.1.1

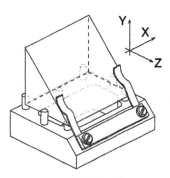

FIGURE 6.1.2

against the contact point of the tertiary plane. Notice that the prism is not precisely positioned along its axis (X). This is permissible because of the prism's cylindrical shape. It is not sensitive to axial position. Therefore, we have decided to *overconstrain* the prism in X. There are *two* contact points constraining the prism in X, one at each end. The space between these two contact points is made to be a little greater than the maximum length prism. Recall from section 1.5 that the penalty for overconstraint is either play or binding. We have opted for play.

The performance test for a demountable connection is to push the body out of contact with each contact point, one at a time, then release it and make sure the nesting force is able to overcome the sliding friction that occurs at all of the other contact points and return the body to its completely nested position. If the body fails to return after any one of its contact points is lifted, a nesting force analysis (section 1.16) will show the reason.

Figure 6.1.3 shows two configurations of demountable connections that are very well known in kinematic circles. They are so well known, in fact, that they are often called "kinematic connections." In each case, the lower body has three rigidly attached spherical features. First, let us examine the configuration shown in Fig. 6.1.3A. The first ball nests in a conical or trihedral socket, providing three constraints through the center of the ball. The second ball nests in a "vee" whose axis intersects the socket. This provides two additional constraints through the center of the second ball and in a plane perpendicular to the groove axis. The degree of freedom remaining at this point is now uniquely defined to be an **R** through the centers of the first two balls. The third ball constrains this degree of freedom by making a single point contact with the upper body. A downward vertical force nests the upper body in place.

The connection in Fig. 6.1.3B is trilaterally symmetrical. The pattern of **C**s consists of three intersecting pairs where the planes of each pair are symmetrically positioned around the perimeter of the parts. Each ball of the lower body nests in a radially oriented groove in the upper body. Let us examine the constraint condition at the instant when just two balls are nested and the third is not. At this instant, each of the two balls will have a pair of constraints passing through its center and lying in a plane perpendicular

FIGURE 6.1.3A

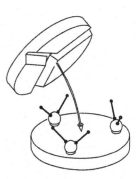

FIGURE 6.1.3B

FIGURE 6.1.4

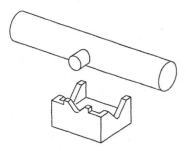

FIGURE 6.1.5

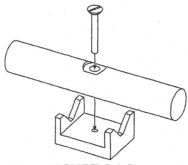

FIGURE 6.1.6

to its associated groove. The total number of constraints is four. We know there must be two degrees of freedom, both **R**s, each intersecting all four **C**s. One **R** is along the line joining the two ball centers. The other **R** is at the intersection of the two constraint planes. Now, if the third ball is permitted to nest in its groove, we can see that two additional **C**s will be applied, removing the two **R**s. Again, the nesting force that maintains all points in contact is vertically downward.

Here is a variation on the configuration of Fig. 6.1.3B, where the balls have been replaced with radially oriented pins and the lower body is a thin ring. Notice that the faces of the vee's in the lower body have been rounded to keep the location of the contact between the pins and the groove faces well defined.

Now consider the effect of repositioning the pins and their associated vees. They do not have to be arranged symmetrically. Figure 6.1.5 shows a demountable connection that uses three "pin-and-vee" connections to precisely attach a body to a round bar. Two of the vees engage the bar itself. With only these features in contact, the body is free to rotate around and translate along the bar axis. The third vee engages a stub post attached to the bar, thus providing tangential and axial constraints.

Figure 6.1.6 shows an idea that is a variation on this configuration. Again, a body is mounted to a round bar using two vees, but this time, a single flat-head screw is used to provide axial and tangential constraint. The screw, which not only supplies the nesting force, also provides tangential and axial constraint through the contact of the conical-shaped underside of its head with the edge of the hole in the body. One must be aware, however, that the precision of position in the axial and tangential directions is diminished owing to the looseness of the screw threads. This connection would not be appropriate for an application demanding very high precision.

The connection between a lens grinding tool and its mounting post is another example of a demountable connection in which the features that provide nesting force also provide constraint. Figure 6.1.7 shows this connection. The post has a cylindrical shape with a rounded end and three ramp features protruding from its cylindrical surface. The grinding tool has a conical socket (which mates with the rounded end of the post) and three internally protruding lugs (which mate with the three ramp features on the post). To assemble, the rounded end of the post enters the socket. The tool is then rotated clockwise (about the post axis), wedging the three lugs against the three ramps.

The contact between the ball and the socket provides three constraints through the center of the ball. The contact at each of the ramps provides three additional constraints, one at each lug, normal to the associated ramp surface. Since the angle of the ramps is less than the friction angle, the connection is "self-holding," like a wedge-shaped doorstop. Thus, the nesting force remains even after the tightening torque has been removed. Figure 6.1.8 summarizes the constraint pattern associated with this connection. Notice that the three Cs at the ramps do not intersect.

Rigid Connection Using Three C-Pairs

We now explore some further uses for the pattern of three intersecting pairs of Cs. In the previous section we saw this pattern implemented with contact points to achieve precision demountable connections. Let us now look for other places where this pattern is used to achieve a precise, rigid connection.

Figure 6.1.9 shows (in exploded view) a pattern of three intersecting C pairs being implemented using sheet flexures as the constraint devices. In this application, the two connected bodies each carry some components of an optical system. The components on body A need to be aligned very accurately with respect to the components on body B. That alignment is carefully done on an accurate fixture in the factory. Once the alignment is completed, the screws are tightened, clamping the flexures in place. After the flexures are securely clamped, the assembly can be removed from the fixture and the critically aligned components remain in position.

I call this connection "mechanical glue," but it is better than glue. It is reusable, does not shrink, has infinite shelf-life, and has infinite "pot life," yet "cures" instantly when the screws are tightened.

The mechanical connection between the space shuttle and the NASA 747 Shuttle Carrier Aircraft uses this same pattern of three intersecting pairs of Cs. Each pair of Cs is implemented by a pair of bars arranged in a "V," one pair under each wing with the "V" lying in a "fore and aft" plane, and the third pair under the nose with the "V" lying in a transverse

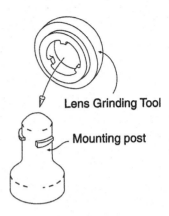

Lens Grinding Tool

Mounting post

FIGURE 6.1.7

FIGURE 6.1.8

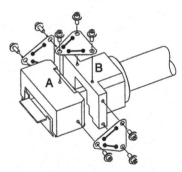

FIGURE 6.1.9

plane. This pattern of **C**s efficiently provides a high strength, low weight, rigid connection between the two vehicles.

It is remarkable that the same pattern of **C**s can be found connecting two giant vehicles or providing submicron precision in the connection between two parts of a laboratory instrument. Clearly, the usefulness of Constraint Pattern Analysis is not bounded by scale.

6.2 ■ FIVE **C**s / ONE **R**

When a body is constrained by a pattern of five **C**s, it is left with exactly one degree of freedom. That degree of freedom is uniquely defined. There can be no more than *one* line in space that intersects each and every one of the five **C**s. That line is the body's single **R**.

Rotors and carriages are alike in that they are connected (to the machine) with five constraints, leaving them with one **R**. The only difference is that with a carriage, the **R** is (usually) located at infinity.

Consider the body shown in Fig. 6.2.1, connected to the bar with two vees spaced apart axially. This body clearly has two degrees of freedom: rotation about the bar axis and translation along the bar axis. By constraining one or the other of these two degrees of freedom, we can obtain either a carriage or a rotor.

By adding a second bar (rail) and a single contact pad, we arrive at this sliding carriage configuration. The additional constraint between the new rail and the pad has removed one of the two degrees of freedom. Only X translation is now allowed.

Alternatively, we can convert one of the V-blocks to a ball-and-socket joint to obtain the rotor configuration of Fig. 6.2.3. The weight of this rotor, acting along the negative Z direction, might provide the needed nesting force.

If we turn the configuration of Fig. 6.2.3 inside-out, we might arrive at the configuration shown in Fig. 6.2.4, where the cylindrical part is the rotor.

To reduce the effect of friction, the contact points have been replaced with wheels. Be aware that the wheels will introduce some imprecision due to runout. To avoid wheel scrubbing, the apex of the cone of each wheel should always coincide with the apex of the conical surface against which the wheel is running. This ensures that at every point of contact, the velocity of the wheel matches that of the surface.

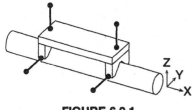

FIGURE 6.2.1

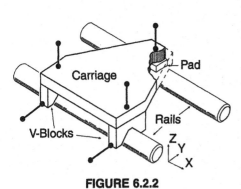

FIGURE 6.2.2

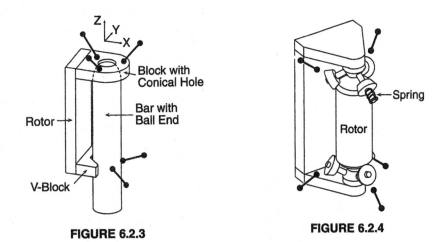

FIGURE 6.2.3

FIGURE 6.2.4

Five of the wheels are mounted in fixed positions. The sixth wheel, which is spring-loaded, provides the nesting force to hold the five constraint wheels in contact. By symmetry, the nesting force at each constraint wheel is equal to the force at the spring-loaded wheel (ignoring gravity).

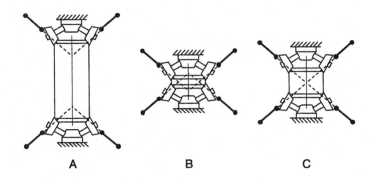

FIGURE 6.2.5

Be careful with this configuration, however. An examination of the constraint pattern points out a potential problem. Figure 6.2.5A shows (in 2D section view) the constraint pattern for a "long" rotor. The single **R** is defined to be the line joining the upper and lower intersection points of the **C**s. Now, look at the constraint pattern for a "short" rotor, shown in Fig. 6.2.5B. Here, the intersection of the lower set of **C**s is *above* the intersection of the upper set. No problem. The single **R** is still defined the same way.

Now consider the situation where we are so unfortunate as to have chosen a "medium" length rotor such as the one shown in Fig. 6.2.5C. Notice that the intersection of the lower set of **C**s is coincident (or nearly so) with the point at which the upper set coincides. This, of course, results in overconstraint of that point while allowing the rotor to be free in *three* rotational degrees of freedom, not just one. This points out the importance of always examining the constraint pattern of a connection to determine that each **C** is positioned at a reasonable distance away from the **R** that it is intended to remove.

The carriage shown in Fig. 6.2.6 is supported by three cables in three degrees of freedom and by three wheels in the remaining three degrees of freedom. The carriage is designed to move vertically in a shaftway. The wheels provide X, Y, and θ_z connection to the walls of the shaftway.

The Z cable, attached to a corner of the carriage and driven by a motor, is used to position the carriage in the Z direction. The other end of the Z cable is attached to an equal counterweight so that the carriage will remain at rest without the application of motor torque.

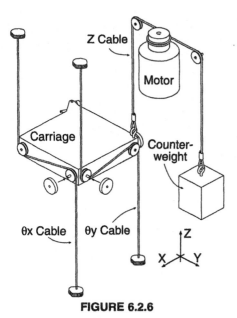

FIGURE 6.2.6

Two other cables constrain the carriage against θ_x and θ_y rotation. The tension in these cables balances the moment exerted by the carriage weight and the tension in the Z cable.

Three wheels connect the carriage to the shaftway in X, Y, and θ_z. A fourth, spring-loaded wheel is positioned in such a way that it loads all three constraint wheels equally.

Figure 6.2.8 shows a one-degree-of-freedom connection used in an autofocusing lens. Two parallel sheet flexures are attached directly (in parallel) to a lens housing. Portions of each sheet flexure have been cut away, leaving just three "spokes" connecting an inner ring and an outer ring. The inner

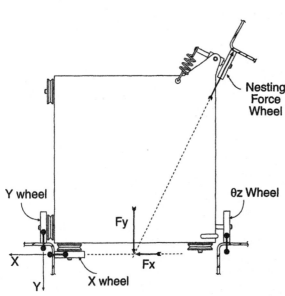

FIGURE 6.2.7

rings attach to the lens housing. The outer rings are fixed to a base part (not shown). Each of the spokes defines a **C**. This pattern of six **C**s is redundant by one, but because of symmetry and because these parts will be positioned and clamped in a special assembly fixture, no adverse effects will result from this over-constraint. The complementary pattern is a single **R** at infinity (at the intersection of the two flexure sheets). This is equivalent to a single translational degree of freedom for the lens to move along its axis.

Figure 6.2.9 shows two configurations that achieve exactly the same constraint pattern as in Fig. 6.2.8 but with an entirely different connection of flexures. In this configuration, six folded sheet flexures are connected directly between a center body and a fixed outer ring, each folded sheet flexure providing a single **C** along the fold. (See section 4.5.) The folds (the **C**s) are arranged to reside in two parallel planes. The Rule of Complementary Patterns reveals the complementary pattern to be a single **R** at infinity, equivalent to a single **Z** translational degree of freedom for the center body.

Figure 6.2.10 shows a flexure connection for a punch and die. The punch is maintained in alignment with the die in five degrees of freedom. The design requirement is that the punch must be allowed to move relative to the die in **Z** but must be constrained in all other degrees of freedom. The nominal clearance between the punch and the die is a fraction of a thousandth of an inch, but the punch must never touch the die.

Because the amount of **Z** motion of the punch is quite small, this motion can be approximated by a small θ_x rotation about a remote axis. That axis is the intersection between the planes of the horizontal and

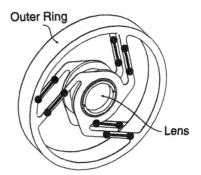

Outer Ring

Lens

FIGURE 6.2.8

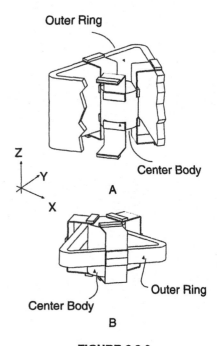

Outer Ring

Center Body

A

Center Body

Outer Ring

B

FIGURE 6.2.9

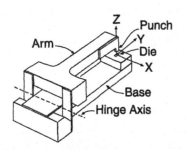

Arm

Punch

Die

Base

Hinge Axis

FIGURE 6.2.10

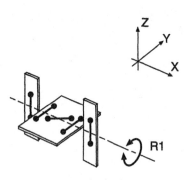

FIGURE 6.2.11

vertical flexures. Figure 6.2.11 shows the constraint pattern of the flexure connection. The only unconstrained degree of freedom is **R**1. The arm is free to rotate (a small amount) about this hinge axis. For small rotations, this is equivalent to a Z motion at the punch.

Figures 6.2.12 and 6.2.13 reveal the actual design of the punch. It embodies some additional features:

- The arm and base are "L" shaped to accommodate the solenoid that "closes" the punch.
- A coil spring (not shown) returns the arm to the "open" position.
- "Open" and "closed" positions are independently screw adjustable. Urethane washers provide cushioned stops.

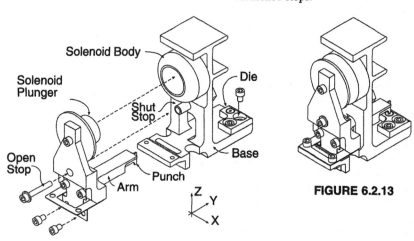

FIGURE 6.2.12   **FIGURE 6.2.13**

- To maximize the stiffness of the punch relative to the die in the X direction, the Y and θ_z constraints need to be spread as wide as possible. For this reason, the outermost flexures have been positioned horizontally while the wide flexure is vertical.
- The solenoid plunger never touches the solenoid body. There is no wear on the solenoid.
- Adjustment is accomplished by loosening the two screws that clamp the die to the base, closing the punch into the die with a thin plastic film to take up the small punch-to-die clearance, and then retightening the two screws.

The problem of differential thermal expansion affecting punch-to-die alignment is avoided by mounting the solenoid on a separate leg.

6.3 ■ FOUR Cs / TWO Rs

Four-C Patterns: Two Intersecting Pairs

If a body is constrained by a pattern of four Cs, positioned in such that they are in two intersecting pairs, that body will have two Rs. Each of those Rs will intersect each and every C. The two lines representing those Rs are easily found:

- One R is the line joining the two intersection points.
- One R is the intersection of the plane of the first pair with that of the second pair.

This is illustrated by the following series of examples showing the connection between a dumbbell-shaped part and the V-grooves in a base part.

For example, in Fig. 6.3.1 the axes of the two V-grooves are parallel. The R shown is the line joining the two intersection points. The second R is at infinity, at the intersection of the two parallel planes, and is represented as a pure translation.

In Fig. 6.3.2, the two V-grooves are colinear. The two degrees of freedom are found the same way.

Fig. 6.3.3, the V-groove axes intersect. Now the planes of the two constraint pairs intersect at a finite location. This intersection defines the location of one of the body's two Rs.

In Fig. 6.3.41, the V-grooves are skew and the dumbbell-shaped part is bent. No matter how twisted or distorted the shape of the parts, the two Rs are easily found simply by knowing the constraint pattern and using the above rule.

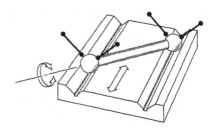

FIGURE 6.3.1

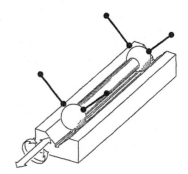

FIGURE 6.3.2

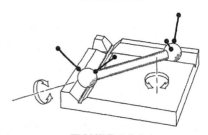

FIGURE 6.3.3

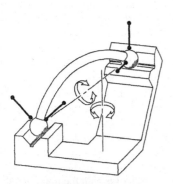

FIGURE 6.3.4

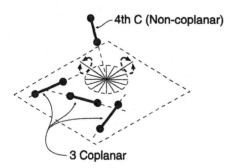

4th C (Non-coplanar)

3 Coplanar

FIGURE 6.3.5

Four-C Patterns: Three Coplanar Plus One

If, in a pattern of four Cs, three are coplanar and the fourth intersects this plane, the complementary pattern of Rs will be two from a disk of radial lines lying in this plane. The center of the disk is at the intersection of the non-coplanar C.

We have seen this constraint pattern before in Fig. 3.4.1, where three of the Cs are coplanar and lie in the plane of the top face of the object. The fourth C intersects that plane at a corner of the object. We saw the object was left with two Rs. Those two Rs are two lines from a disk of lines lying in the plane of C_1, C_2, and C_3 and intersecting C_4. The body's two Rs are not *uniquely* determined. Any two lines from that disk will intersect every one of the four Cs.

Two Rs Define a <u>Ruled Surface</u>

Whenever a body is constrained by a pattern of four Cs, that body will have two degrees of freedom. The pattern of four Cs defines a complementary pattern of two Rs. Those two Rs define a "ruled" surface containing an infinite number of (infinitely long) straight lines, any two of which can be considered to represent the body's two degrees of freedom.

If the two Rs intersect, then the ruled surface is a planar disk of radial lines containing the two defining Rs (see section 3.4). Any two lines from this disk, provided the angle between them does not approach 0 degrees (or 180 degrees), can be considered to represent the body's two Rs.

If the Rs are parallel, then the ruled surface is a flat plane of parallel lines containing the two defining Rs (see section 3.5). Any two lines from the plane, provided they are not spaced very close together (in relation to the dimensions of the body) can be considered to represent the body's two Rs.

If the two Rs are skew (neither intersecting nor parallel), then the ruled surface is a cylindroid that contains the two defining Rs and whose axis is the line mutually perpendicular to the two defining Rs (see section 3.10). Every line on this cylindroid represents a helical degree of freedom (screw) of a certain pitch. The body's two degrees of freedom can be considered to be represented by any two lines from this surface, provided the lines are not chosen to be too close together (as with the plane and the disk).

When a body constrained by a pattern of five Cs was found to have one R, we found that there was a *unique* location for that **R**. There was never more than one line in space found to intersect every line of a pattern of five nonredundant lines.

When a body is constrained by a pattern of four **C**s, we know that it has two degrees of freedom, but these two degrees of freedom are not uniquely positioned in space. They lie on a ruled surface that contains an infinite number of lines. The body's two degrees of freedom are two lines from this surface.

Approaching a Design Problem

Suppose you need to design a connection that has two degrees of freedom. You might approach this problem in a few different ways.

1. Knowing the location of the two desired **R**s, **R**1 and **R**2, simply design a *cascaded* connection with two hinges.
   a. If the two **R**s intersect, consider the possibility of positioning the hinges along the "best" two radial lines of the disk defined by **R**1 and **R**2.
2. Knowing the location of the two desired **R**s, find the complementary pattern of four **C**s, then design a *direct* connection to provide this pattern.
   a. If the **R**s intersect, define the complementary pattern to be three coplanar **C**s plus a single **C** through this plane at the intersection of the **R**s.
   b. If the **R**s do not intersect, define the complementary pattern to be two intersecting pairs of **C**s.
   c. Rearrange the **C** pattern further by using the rules of **R**s and **C**s.

Some familiar examples will illustrate these solution paths. Suppose a roller needs to be mounted so that it is free to rotate through small excursions about axes **R**1 and **R**2, as shown in Fig. 6.3.6.

The simplest and most obvious solution is to design two cascaded hinges which might look something like Fig. 6.3.7. A designer does not need to know anything about Constraint Pattern Analysis techniques to come up with this solution. This is solution path 1.

But now suppose that there is a restriction that the hinge hardware cannot be located as shown in Fig. 6.3.7, but must be placed *above* the device instead. Using our knowledge of the equivalence of all radial lines defined by **R**1 and **R**2, we realize that

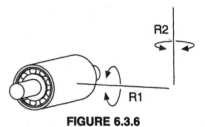

FIGURE 6.3.6

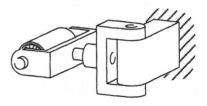

FIGURE 6.3.7

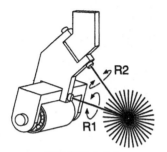

FIGURE 6.3.8

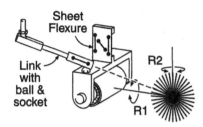

FIGURE 6.3.9

R1 and **R**2 will still be achieved by the apparatus of Fig. 6.3.8. This is solution path 1a.

Now, let us consider the possibility of designing a direct connection to the device using a complementary pattern of four **C**s. Solution path 2a suggests the connection illustrated in Fig. 6.3.9. The sheet flexure lies in the plane of **R**1 and **R**2 and contains three coplanar **C**s. The fourth **C**, imposed by the link with ball-and-socket end connections, cuts through the plane of the flexure right at the location of the intersection of **R**1 and **R**2.

Another example illustrates solution path 2b. Figure 6.3.10 shows a lens that must move in two degrees of freedom, **Z** (focus) and **X** (tracking), as on a compact disk player.

It is acceptable to approximate the **X** freedom with remotely positioned rotational freedom, **R**1. The **Z** freedom is expressed as **R**2 at infinity.

The four-**C** pattern shown in Fig. 6.3.12 is complementary to the two-**R** pattern. The **C**s are arranged in two intersecting pairs. **R**1 is the line joining the two intersection points. **R**2 is the intersection of the planes of each pair.

Figure 6.3.13 shows four wire flexures being used to implement the four-**C** pattern found in Fig. 6.3.12.

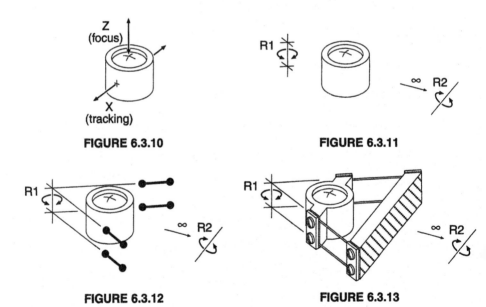

FIGURE 6.3.10

FIGURE 6.3.11

FIGURE 6.3.12

FIGURE 6.3.13

6.4 ■ THREE Cs / THREE Rs

The device pictured in Fig. 6.4.1 is a castered and gimbaled tension roller. As implied by its name, it has three degrees of freedom: (1) caster, (2) gimbal, and (3) tension. (Actually, it has a fourth, roll, but we will ignore this for now.) These three degrees of freedom are accomplished by a cascaded connection from the base to the carriage (X), from the carriage to the rotor (θ_x), and from the rotor to the yoke (θ_z).

There is quite a bit of hardware used to accomplish these three degrees of freedom for the yoke. Can we simplify this mass of hardware by replacing it with a direct connection to accomplish the same three degrees of freedom?

The three desired **R**s are shown in Fig. 6.4.2, in relation to the yoke. The direct connection would consist of three **C**s that are the complement of the three desired **R**s.

The complementary pattern would consist of three nonredundant lines selected from the two ruled surfaces shown in Fig. 6.4.3. One surface is the vertical plane of vertical lines that goes through the intersec-

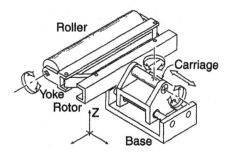

FIGURE 6.4.1

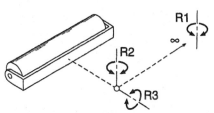

FIGURE 6.4.2

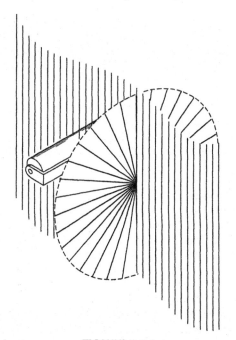

FIGURE 6.4.3

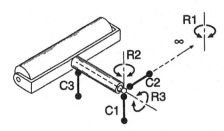

FIGURE 6.4.4

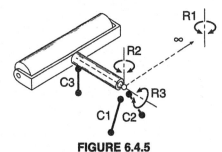

FIGURE 6.4.5

FIGURE 6.4.6

tion point of $R2$ and $R3$, parallel to $R1$. The second surface is the vertical disk of radial lines, whose center is at the intersection of $R2$ and $R3$ and which contains $R1$ and $R2$.

Two of the C lines would be selected from one surface, and one from the other. To use Cs from both surfaces (attach them to the yoke) it is clear that we must add some sort of "stem" to the yoke, as in Fig. 6.4.4. Here, $C1$ is along the intersection of the two surfaces. $C2$ is from the disk and $C3$ is from the plane of parallel lines.

Now, assuming that we wish to make our connection to a base that is in the same relative position as the base in Fig. 6.4.1, we might instead choose the three Cs shown in Fig. 6.4.5. Here, $C1$ and $C2$ are from the disk and $C3$ is from the plane of parallel lines. The resulting degrees of freedom, of course, are the same.

The hardware to implement this C pattern might look as shown in Fig. 6.4.6. This hardware is certainly much simpler and cheaper than that shown in Fig. 6.4.1, yet both accomplish the same result: exactly the same three degrees of freedom for the yoke.

> The greater the number of degrees of freedom, the greater the potential improvement (in simplicity) in using a direct connection rather than a cascaded connection.

Now go back to Fig. 6.4.3. There are two surfaces, each containing an infinite number of lines, from which to select our three Cs.

We made that selection according to our convenience. We chose lines that went from the yoke to the base. Now, let us work this problem in reverse. Suppose we start with our three Cs and find "the complementary pattern of Rs." We see that the three Rs we had before ($R1$, $R2$, and $R3$) are not a unique answer. They are just three nonredundant lines from the two ruled surfaces shown in Fig. 6.4.7.

We now see a trend. When we had five Cs and one R, the position of that R was uniquely determined. When we had four Cs and two Rs, those two Rs were *not* uniquely determined, but were found to reside among the infinite number of lines on a ruled surface. Now with three Cs and three Rs, we find that those three Rs are confined to reside among the

lines of two ruled surfaces. Furthermore, the complementary pattern of three **C**s was also found to reside
among the lines of two ruled surfaces. The two ruled
surfaces on which the **C**s are resident are "complementary" to the two ruled surfaces on which the **R**s
are resident. Look closely at Figs. 6.4.3 and 6.4.7 to
see what is meant when we say the surfaces of one
figure are the complement of the surfaces of the other
figure. *Any line* that resides on either ruled surface of
one figure intersects *every line* that resides on both
ruled surfaces of the other figure.

The condition of three **C**s / three **R**s is symmetrical. We have just seen one example of that symmetry.
The disk and plane of Fig. 6.4.3 are complementary
to the plane and disk of Fig. 6.4.7.

We have seen some other examples of the symmetry that occurs between complementary patterns of
three **C**s / three **R**s. For example, recall the sheet
flexure. It imposes a pattern of three coplanar **C**s.
The complementary pattern is three coplanar **R**s in
the same plane.

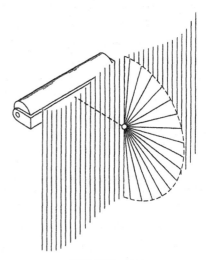

FIGURE 6.4.7

A pattern of three nonredundant coplanar lines
defines a complementary pattern that is three
nonredundant lines from the set of all lines resident in that plane.

Now recall the ball-and-socket joint, which
imposes a pattern of three **C**s intersecting at the center of the ball. The complementary pattern is three **R**s
also intersecting at the ball center.

A pattern of three nonredundant lines intersecting
at one point defines a complementary pattern that
is three nonredundant lines from the set of all lines
intersecting that point.

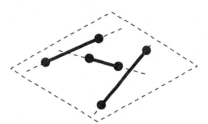

FIGURE 6.4.8

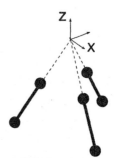

FIGURE 6.4.9

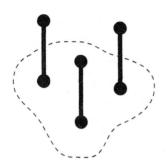

FIGURE 6.4.10

FIGURE 6.4.11

Now consider the case where the intersection point is at infinity. In this case, the three lines are parallel.

> A pattern of three nonredundant parallel lines defines a complementary pattern of three nonredundant lines from the set of all lines parallel to the three starting lines.

This is exactly the situation we had in the 2D case throughout Chapter 1. To have 2D, we tacitly imposed three parallel Z direction constraints at the outset. This left our "unconstrained" 2D model with three $\mathbf{R}$s, each residing among a "bundle" of parallel lines, all lines parallel to the Z-axis. When we applied our first constraint to the 2D model, we were left with two degrees of freedom, two $\mathbf{R}$s from the ruled plane of parallel lines that intersected that $\mathbf{C}$. The second applied $\mathbf{C}$ then removed all but one $\mathbf{R}$. This single, uniquely positioned $\mathbf{R}$ was at the intersection of the two $\mathbf{C}$s (and parallel to the three Z-direction $\mathbf{C}$s that were tacitly applied at the outset to define the 2D problem).

There is one more symmetrical configuration of three $\mathbf{C}$s / three $\mathbf{R}$s that should be mentioned before leaving the topic: three skew lines that do not intersect. Suppose a cubical body is constrained by a pattern of three $\mathbf{C}$s as shown in Fig. 6.4.11. These lines are skew. They do not intersect each other.

If we were to find every line in space that intersects all three of those $\mathbf{C}$s, we would get the collection of lines shown in Fig. 6.4.12. These lines are generators of a continuous, ruled surface that is called a hyperboloid. The hyperboloid actually has two sets of generators. One set of generators, the "left-handed" set, is the set of all lines that we just found. The other set of generators, the "right-handed" set, also lies in the surface of the same hyperboloid. Each right-handed generator intersects every left-handed generator.

The three $\mathbf{C}$s that defined our problem are members of the right-handed set. Any three lines from the left-handed set can be considered to represent the body's three $\mathbf{R}$s. The three $\mathbf{C}$s applied to the body could have been positioned along any three lines of the right-handed set and provided an equivalent constraint condition. The resulting pattern of $\mathbf{R}$s would have been the same. (As usual, of course, the three lines chosen need to be judiciously spaced to avoid redundancy.)

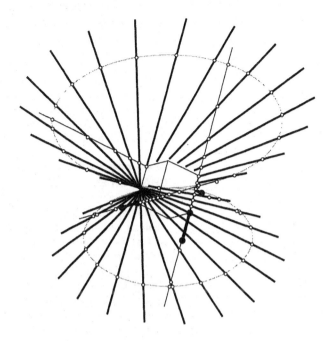

FIGURE 6.4.12

6.5 ■ RECIPROCAL PATTERNS

We do not need to explore patterns of four, five, and six **R**s and their complementary patterns of **C**s. The Constraint Pattern Analysis techniques for four **R**s / two **C**s are just the same as starting with a pattern of four **C**s and looking for the complementary pattern of two **R**s. For example, in section 6.3, we saw several patterns consisting of four **C**s and then found the complementary pattern of two **R**s.

It is, therefore, not necessary to describe exhaustively all possible patterns of four **R**s and their complementary patterns of two **C**s. It is only necessary that the reader become fluent in the techniques of Constraint Pattern Analysis so that when faced with a pattern of four lines, he or she can quickly find the complementary pattern of two lines. It is my hope that the reader has by now gotten a strong sense of the symmetry that exists between **R**s and **C**s in general, and also of the symmetry that exists around the number 3. For example, when we have three nonredundant lines, the complementary pattern of three lines can be found whether we start with **R**s and find **C**s or start with **C**s and find **R**s.

Now we see the same holds true with patterns of four/two. If we start with four nonredundant lines, the complementary pattern of two lines can be found whether we are going from **R**s to **C**s or from **C**s to **R**s. In other words, if we are able to find the complementary pattern of two **R**s, given a pattern of four **C**s, then we have already solved the reciprocal patterns of four **R**s and two **C**s.

CHAPTER SUMMARY

In this chapter, many more examples were presented illustrating the application of Constraint Pattern Analysis techniques to various familiar hardware configurations. Although the mechanical connections reviewed in this chapter are organized by the number of **C**s applied, it does not attempt to be a complete catalog of connections. It does, however, present some interesting ideas. For example, in the section on connections having six constraints (zero freedoms), various precision demountable connections are presented. These are connections that can be easily taken apart and reassembled into a precise, repeatable position. Such connections have great utility in precision optical apparatus, for example, where a component might need to be removed periodically for cleaning then replaced in an exactly repeatable position. In the section on connections having five constraints (one freedom), some important details on the constraint pattern configuration of a rotor are revealed. Several interesting examples presented in the four constraints (two freedoms) section illustrate the use of some handy techniques for finding the complementary pattern of two lines, given a starting pattern of four lines. Also, a systematic process for design synthesis is suggested and illustrated with hardware examples. Finally, in the section on three constraints (three freedoms), the idea of complementary patterns of three lines is summarized and the idea of complementary surfaces is introduced.

CHAPTER 7

Structures

Many mechanical design engineers are accustomed to thinking of structures as being categorically different from mechanisms. After all, mechanisms have moving parts, whereas structures are inanimate.

But here we examine structures as if they are no different from mechanisms. Certainly we can imagine constraining a mechanism so it has no degrees of freedom. Does this make it a structure? We are also about to discover that it is all too easy to botch the design of a structure so that it has one or more unwanted degrees of freedom. Does this make it a mechanism?

7.1 ■ INTRODUCTION

In an informal paper titled "Strong, Lightweight Structures," written in the 1960s by Dr. John McLeod of the Eastman Kodak Company, the reader is asked "How many legs does a good tripod have?"

To find the correct answer to this seemingly simple question requires one to understand some of the most fundamental principles of kinematic design, concepts that some might regard as common sense or just sound mechanical design practice. But Dr. McLeod points out that most mechanical designers and engineers are unable to answer correctly. In fact, even some authors on the subject of precision instrument design have also given the wrong answer.

Obviously, the purpose of a tripod is to provide rigid positioning for a platform to which an instrument (e.g., a surveyor's instrument or camera) will be mounted. For this purpose, we know that the instrument platform must be constrained against motion in six degrees of freedom, so there must be six constraints between the instrument platform and the ground. These constraints might be arranged in a pattern such as that shown in Fig. 7.1.1. You may recognize this pattern of constraints to be similar to the pattern of constraints used in the demountable connections illustrated in Fig. 6.1.3B and Fig. 6.1.4. This pattern of constraints clearly provides a rigid connection to the platform in which all six of the platform's degrees of freedom are exactly constrained.

Next, each constraint can be replaced by an "equivalent bar" or leg. (The legs are longer than the constraint symbols of Fig. 7.1.1 so they come together in pairs at the bottom.) When we replace each constraint of Fig. 7.1.1 with an equivalent leg to arrive at the configuration of Fig. 7.1.2, we acknowledge the "binary constraint" nature of the long, slender legs. Like a wire flexure, a long slender leg or bar has high axial stiffness compared with its bending stiffness. To provide optimal stiffness to the instrument platform, we design as if the legs were connected with ball joints at both ends. The fact that the legs are not actually connected with ball joints simply serves to supplement the stiffness of our structure still further. The small contribution of the leg's bending stiffness is added to the predominant axial stiffness of the legs.

It is now clear that the proper number of legs for a tripod is six! The reason it is called a tripod is that

FIGURE 7.1.1

FIGURE 7.1.2

it has three feet. The legs happen to be arranged so that they join in pairs at the feet. Unfortunately, the definition of "tripod" has, over the years, broadened to include not only three-footed structures, but also three-legged structures as well. The distinction between legs and feet has become muddled. Thus, the term "tripod" has become ambiguous. But, when we design structures, we certainly cannot afford to be ambiguous.

A long slender leg can provide only one constraint (along its axis). The term "foot" refers to the end of a leg. More than one leg might have a common foot, as in the case of the instrument tripod.

Actually, there are some cheap tripods that have only three legs, instead of six. They rely on the bending stiffness of the legs for the three rotational degrees of freedom of the instrument. But, performance-wise, the rotational degrees of freedom of the instrument are often even more important than the translational degrees of freedom, so these three-legged tripods can be expected to give especially poor performance compared with the six-legged ones.

The lesson of the tripod is useful because the six-legged connection between two rigid bodies shows up in lots of other places as well. For example, in connecting the space shuttle to its huge liquid fuel tank during launch, and to the top of a specially modified Boeing 747 for transcontinental flights, a similar six-legged connection is made. Obviously, in those connections, light weight and high strength and stiffness are very important. Another place where this type of connection is used is in the mounting of aircraft and rocket engines. In this application, the ball joints at the ends of the bars are not just conceptual, they are really needed to avoid bending stresses that would occur as the engine's dimensions change during warm-up. The joints at the bar ends must actually undergo some rotation as the engine temperature changes. In this situation, overconstraint would be a real "no-no." No attempt is made to use the bar's bending stiffness to supplement axial stiffness as in the tripod.

Similar precautions need to be taken in the design of structures to mount precision instruments. If the connection between the instrument and the "outside world" is overconstrained, then any distortion of the outside world will produce unwanted stresses and distortion of the instrument.

We generally think of a structure as an arrangement of fixed mechanical members (e.g., plates and bars) whose purpose is to provide rigid connection between objects, usually in six degrees of freedom. These plates and bars are the constraint devices of a structure.

The bars of a bar structure should be thought of in the same way that we think of wire flexures, extremely stiff in supporting straight axial loads, but flexible in response to bending loads. The sheets and plates of cast and molded structures should be treated just like sheet flexures, unable to support loads that do not lie in the plane of the sheet, but very stiff in resisting in-plane loads.

Recall that with flexures, the ratio between bending and stretching stiffness of sheets and wires was so enormous that it did not require a very big leap of faith to completely ignore bending stiffness. We used ball joints to idealize this lack of bending stiffness. We conceptually replaced wire flexures with bars having ball-joint ends. Each bar was considered to provide a single axial constraint. Similarly, we identified "equivalent" bars in sheet flexures. We found the equivalent bar structure to consist of three bars with ball-and-socket ends.

In a structure, however, it is common to have members (bars and plates) whose thickness-to-length ratio is greater than that of a flexure. For this reason, those members have bending stiffness that is more appreciable than it is with flexures. Nevertheless, the stiffness of these bars and plates in stretching will always be far greater than their stiffness in bending. Therefore, it can be concluded:

> When designing a structure for optimal stiffness, it is important that all of its members (bars and plates) be used in stretching or compressing rather than bending.

Therefore, as a conceptual tool, it is still advantageous to imagine that the material of a structure is composed of equivalent bars whose ends are connected with ball joints. This is exactly the technique that we used in designing and analyzing flexures, and it applies to structures as well. Structures should be thought of as being composed of flexure-like members.

It is ironic (and unfortunate) that most designers regard structure design as mundane (perhaps because structures are inanimate), but they think of flexure mechanisms as sophisticated, for use in high-tech applications where extreme precision is required. In truth, structures require the same constraint pattern analysis as flexures. Structures and flexure mechanisms are categorically similar. The fundamental principles of structure design are exactly the same as those for flexure design, and only after the principles of flexure design are well understood does it make sense to study structure design.

7.2 ■ RIGID SHAPE VERSUS FLEXIBLE SHAPE

Consider the simple beam of Fig. 7.2.1 supporting load P. The beam will *bend* under the influence of load P such that its midpoint will deflect an amount, δ.

Now imagine that the same material of this beam has been reconfigured to the triangular geometry shown in Fig. 7.2.2. The application of load P' to this triangular truss results in a very much *smaller* midspan deflection, δ'. This is because the forces are now carried along the *length* of the bars, trying to *stretch* or *compress* them, whereas in the case of the beam, force P acted to *bend* the bar. This illustrates the difference between a *rigid-* and *flexible-*shaped structure.

> Loads applied to a *rigid* structure result in forces being carried along the length of its members, causing the members to be loaded in *tension* or *compression*, not in bending. Loads applied to a *flexible* structure result in *bending* deflection.

I use the term *rigidity* as a qualitative property. When a structure is a *rigid* structure, it is rigid by virtue of its shape. It is extremely stiff compared with a structure (of the same material) that has a *flexible* shape.

7.3 ■ SHEETS

In our study of flexures, we learned that when a sheet of material is used to connect objects, it provides exactly the same constraint pattern as if three bars had been used. When sheets or plates are used as structural members, they can likewise be thought of

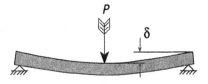

FIGURE 7.2.1

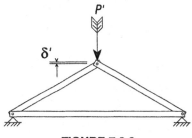

FIGURE 7.2.2

in terms of "equivalent bars" lying within the plane of the sheet material. Such a pattern of equivalent bars is arranged to divide the area of the sheet into triangular apertures. Throughout this chapter, we frequently and freely convert back and forth between sheet structures and their equivalent bar structures.

Flanges

When sheet metal is used in a structure, it is generally considered good practice to bend a flange along the edge. The reason for this is that the flange increases the cross-sectional moment of inertia of the equivalent bar lying along that edge, allowing it to withstand a larger compressive load without buckling.

7.4 ■ "IDEAL" BARS

Because the bars contained in a rigid structure are all loaded axially, never in bending, it is helpful to imagine that the bars of a structure are all "ideal" bars, connected at their ends by perfect ball-and-socket joints, even if we plan to weld the structure in actual practice. Such an ideal bar would possess *one-dimensional rigidity* along its axis. It would rigidly support loads only along its axis. Loads applied at the joints of a bar structure can be broken into components that lie along the axes of the individual bars.

7.5 ■ 2D RIGIDITY

Whereas a bar possesses one-dimensional rigidity, a flat sheet of material such as metal, or even paper, is a good example of a *two-dimensionally rigid* object.

A body that possesses 2D rigidity will rigidly resist 2D distortion or shape change. For example, a rectangular 3×5 card will rigidly resist being deflected into a parallelogram shape by the in-plane shear force P.

The reason for this is clear when we examine the equivalent bar pattern of the rectangular sheet. We are familiar with this structure as a "fence gate." The diagonal bar imparts 2D rigidity by partitioning the rectangle into two rigid triangles. The triangle is the minimum bar structure having 2D rigidity.

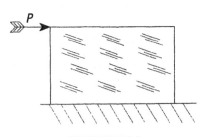

FIGURE 7.5.1

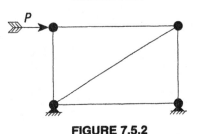

FIGURE 7.5.2

7.6 ■ 2D JOINT ADDITION

It is possible to construct any number of different 2D rigid trusses by a process known as 2D joint addition. In this process, a new joint is added to a rigid starting structure by adding two new bars. The new joint is connected by two new bars to two different joints on the starting structure. (Of course, to avoid overconstraint, the angle between the two new bars must not be close to 0 degrees or 180 degrees.) For example, we could use the process of joint addition to construct the fence gate structure from a starting triangle.

This process can also be used to design large, rigid 2D trusses such as the ones shown in Fig. 7.6.1. In each of these trusses, the number of bars is related to the number of joints by the following formula:

$$B = 2J - 3 \qquad (1)$$

where B is the number of bars and J is the number of joints

This will hold true for any (2D) truss that has been derived from a triangular starting structure using the process of joint addition. Such a truss is said to be statically determinate or *exactly constrained.*

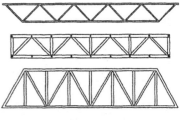

FIGURE 7.6.1

7.7 ■ UNDERCONSTRAINT AND OVERCONSTRAINT IN STRUCTURES

It is possible to have the wrong number of bars (in relation to the number of joints) in a 2D structure.

Too few bars in a structure results in *underconstraint*. In the underconstrained structure shown in Fig. 7.7.1, there are four bars and four joints. The proper number of bars for an exactly constrained truss can be determined (from Eq. 1) to be five.

Because the structure is missing one bar, it is said to be *underconstrained* in one degree of freedom.

On the other hand, too many bars in a structure results in *overconstraint*. The structure shown here is overconstrained in one degree of freedom. The penalties associated with overconstraint in a bar structure are exactly the same as enumerated in section 1.5. To install the final (sixth) bar, its length must *exactly* match the corresponding dimension of the structure or it will not fit. This can be achieved either by holding tight tolerances or by special assembly techniques, such as drilling in place during assembly.

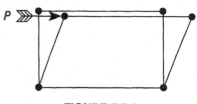

FIGURE 7.7.1

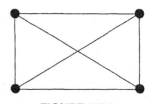

FIGURE 7.7.2

Then, once the structure is completely assembled, a length variation in any of the bars (such as might be caused by a nonuniform temperature change) will result in the buildup of internal stress.

It should be noted that sheets and plates, by definition, begin overconstrained. A sheet "contains" an infinite number of "equivalent bars," which lie everywhere throughout the plane of the sheets.

There is, however, one advantage of overconstraint: it results in a structure with greater stiffness. For this reason, overconstraint in a structure can be advantageous.

7.8 ■ 3D RIGIDITY

The structures that are the most useful to us in the design of machines are three-dimensionally rigid structures. These will rigidly resist deflection due to loads in any direction, not just loads applied within a plane as with two-dimensionally rigid structures.

To appreciate this distinction, let us look again at the fence gate structure of Fig. 7.5.2, which we found to be two-dimensionally rigid. We first consider what will happen when we subject this structure to the load pattern shown in Fig. 7.8.1, which consists of two opposed couples trying to *twist* the structure.

Clearly, it will deflect. It can be thought of as two triangles that are free to pivot relative to each other about a hinge that lies along their common bar. Although this structure is two-dimensionally rigid, it lacks three-dimensional rigidity, which we now define as follows:

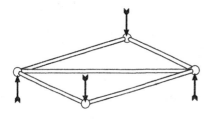

FIGURE 7.8.1

> A three-dimensionally rigid structure rigidly resists deflection by torsional loads by carrying the applied loads along the axes of the structure's bars and in the plane of the structure's sheets.

It is clear that a 2D (planar) structure is incapable of being three-dimensionally rigid. Only a three-dimensional structure can be three-dimensionally rigid.

7.9 ■ 3D JOINT ADDITION

A useful technique for synthesizing custom three-dimensionally rigid structures is to use a process known as 3D joint addition. In 3D joint addition, a new joint is added to a rigid starting structure by connecting the new joint with three new bars to three different joints on the starting structure. The three new bars must, of course, be non-coplanar in order to exactly constrain the new point to the starting structure. As an example of 3D joint addition, we can add a joint to a triangular bar structure to arrive at a tetrahedron.

The tetrahedron is the simplest of three-dimensional structures. It is also the simplest of three-dimensionally *rigid* structures. When loads are applied to the corners of a tetrahedron, those loads are resolved into component forces that are carried by the structure's bars as pure tension or compression loads.

By again using the process of 3D joint addition, we can add a joint to the tetrahedron to arrive at the structure of Fig. 7.9.2.

This structure contains the fence gate plus four additional bars joining at an out-of-plane point. This is a three-dimensionally rigid structure.

Whenever joint addition is used to modify three-dimensionally rigid starting structure, the resulting structure will also be three-dimensionally rigid. For three-dimensionally rigid bar structures, the number of bars will always be related to the number of joints by the following formula:

$$B = 3J - 6 \qquad (2)$$

where B is the number of bars and J is the number of joints

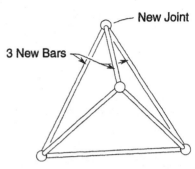

FIGURE 7.9.1

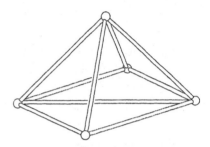

FIGURE 7.9.2

7.10 ■ THE SIX-SIDED BOX

The six-sided box is probably the most commonly occurring three-dimensionally rigid shape in structures. It can be derived from the tetrahedron bar structure using principles we have already learned. Start with the tetrahedron bar structure shown in Fig. 7.9.1. Next, use 3D joint addition to add four new joints as shown in Fig. 7.10.1. The resulting structure is the "bar equivalent" of a six-sided box. The twelve new bars represent the edges of the box. The six bars of the starting tetrahedron are the face diagonals. The sheet equivalent of this bar structure is a closed, six-sided box.

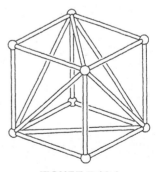

FIGURE 7.10.1

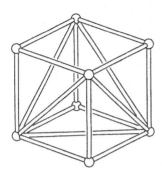

FIGURE 7.11.1

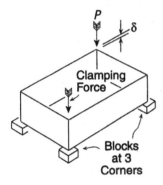

FIGURE 7.11.2

7.11 ■ COMPARING A FIVE-SIDED BOX WITH A SIX-SIDED BOX

As an experiment, we remove the top face diagonal from the structure shown in Fig. 7.10.1 to arrive at Fig. 7.11.1. The resulting structure is the bar equivalent of a five-sided box.

Now, using the test apparatus shown in Fig. 7.11.2, compare the torsional stiffness of a five-sided box with that of a six-sided box. The box is supported at three corners. The fourth, unsupported corner, is subjected to load *P*. Deflection δ is measured. The ratio P/δ gives an indication of torsional stiffness. Indeed, when this test is carried out for both the five-sided box and the six-sided box, regardless of whether the boxes are constructed of cardboard or metal, the stiffness of the six-sided box is *orders of magnitude* higher than the stiffness of the five-sided box. Now recall that the equivalent bar structures of the (closed) six-sided box and the (open) five-sided box differ by only a single diagonal bar across the top face. Because the five-sided box is missing exactly one bar compared with the three-dimensionally rigid six-sided box, the five-sided box is said to be *underconstrained* in one degree of freedom.

That degree of freedom could be described in a few different ways. For example, we could say that the diagonal distance across the open face is unconstrained, permitting the open face to become parallelogram-shaped; or we might notice that opposite sides of the box rotate relative to each other; or we could note that each of the sides and the bottom become warped (twisted). Regardless of how we describe this degree of freedom, the structure has just one degree of freedom. To remove this degree of freedom and restore the structure's three-dimensional rigidity, we must restore the two-dimensional rigidity of the open face by replacing the missing diagonal bar. In order for the box (made of thin sheet material) to be three-dimensionally rigid, each face must be two-dimensionally rigid.

As a practical matter, it is very seldom that a completely closed box can be used as a rigid structure in a machine. How would we gain access to the components inside? An open box would be much more useful from the standpoint of access to the inside. Therefore, we need to consider ways of "structurally closing" open faces of a shell enclosure. In other words, we are looking for equivalent structures that will render an open face two-dimensionally rigid.

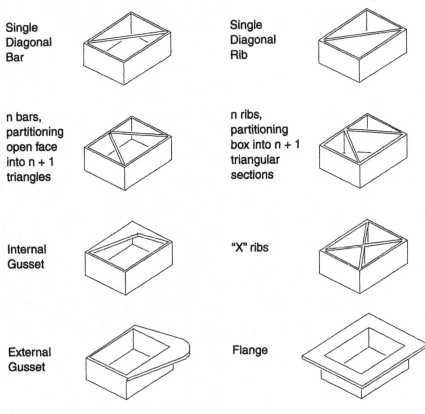

Single Diagonal Bar

Single Diagonal Rib

n bars, partitioning open face into n + 1 triangles

n ribs, partitioning box into n + 1 triangular sections

Internal Gusset

"X" ribs

External Gusset

Flange

FIGURE 7.11.3

Figure 7.11.3 shows several alternative ways of structurally closing the open face of a five-sided box.

In Fig. 7.11.4, holes of various shapes have been cut into several faces of this six-sided box. The box is still three-dimensionally rigid because each face is structurally closed. The material remaining in each face is sufficient to act as an internal gusset, ensuring that the 2D rigidity of each face is maintained.

Figure 7.11.5 demonstrates that the holes can even cut through the edges of the box without destroying the box's 3D rigidity, as long as each face is still two-dimensionally rigid.

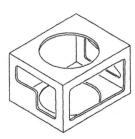

FIGURE 7.11.4

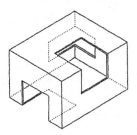

FIGURE 7.11.5

7.12 ■ POLYHEDRAL SHELLS

Using the process of 3D joint addition, we have now devised three different three-dimensionally rigid bar structures. These three bar structures (the tetrahedron, the pyramid, and the box) have equivalent sheet structure counterparts that are also rigid. Because the faces of each of these bar structures are two-dimensionally rigid (the faces contain diagonal bars partitioning the face into triangles), each of these faces can be replaced by an equivalent sheet.

The thing that each of these structures has in common is that each is a closed polyhedral shell (or bar equivalent). This turns out to be the common denominator of all three-dimensionally rigid structures. Just as the sheet was found to be the minimum 2D rigid shape, we now discover that the structurally closed polyhedral shell is the minimum 3D rigid shape.

> The minimum structure required for three-dimensional rigidity is a polyhedral shell or the bar equivalent structure. To be three-dimensionally rigid, the polyhedral shell must be *structurally closed*. This means that each and every face of the polyhedral shell must be two-dimensionally rigid.

Tubes

Many of the polyhedral shells that we encounter in the design of structures for instruments and machines fall into the category of *tubes*. We generally think of a tube as a cylindrical shell of some arbitrary cross-sectional shape (not necessarily round). To be three-dimensionally rigid, a tube must have structurally closed ends. "Structurally closed" means that the tube's cross-sectional shape is constrained against distortion.

If the end faces are sheets, this requirement is met automatically. However, if bar construction is used, each face must be partitioned into triangles. For example, a quadrilateral tube with *open* ends would be underconstrained in two degrees of freedom, as the cross-sectional shape of each end could be distorted. To constrain a quadrilateral against two-dimensional distortion, a single diagonal bar is needed. A single diagonal bar will partition the quadrilateral into two triangles. In general, the number of bars needed to "close" any polygon is three less than the number of sides of the polygon ($B = N - 3$). Note that a triangle is intrinsically closed. There is no need to close the ends of a triangular tube.

Now consider a hexagon. Three bars are required to close this shape. The bars might be arranged as shown in Fig. 7.12.1.

If we had a hexagonal tube, three bars would be required at each end to achieve structural closure. If only one end of the tube was closed, the structure would be underconstrained in three degrees of freedom. The number of missing bars equals the degree of underconstraint.

Applications

Structural rigidity is very important in the design of structures for instruments. The extremely high stiffness allows the instrument to undergo only infinitesimally small deflections under the application of normal operating forces and loads. Furthermore, the high stiffness-to-weight ratio of a rigid structural shape causes the natural frequencies to be as high as possible and, therefore, amplitude of vibration to be as low as possible.

But instruments are not the only place where optimum stiffness and, therefore, rigid shapes are needed. Large structures such as highway signs, towers, and excavating machinery are just a few examples of applications where rigid shapes are used because they can efficiently withstand the large forces to which they are subjected. In each of these cases, we can see tubes or polyhedral shells being used to provide three-dimensional rigidity.

This highway sign is supported by a bar structure in the form of a triangular tube.

High-voltage electric power transmission wires are supported on towers that typically have a square tubular shape, tapering aloft.

The booms of the shovel in Fig. 7.12.4 are of a tubular construction. Note that the cross-sectional size and shape vary along the length. Also, in one tube, the axis is bent.

The list goes on and on. Tubular structures are common in aerospace vehicles, in sporting equipment (tennis racquets, skis, surfboards, etc.), in machine tools, optical fabrication tools and equipment, and in construction equipment. It seems remarkable that the direction of goodness in structure design is the same for such a diverse array of purposes and for such a broad scale of sizes.

FIGURE 7.12.1

FIGURE 7.12.2

FIGURE 7.12.3

FIGURE 7.12.4

7.13 ■ LOCUS OF RIGIDITY

When we make connections to a structure, we open a path by which loads can be applied to the structure. Therefore, we need to ensure that those connections are made to points on the structure that are capable of rigidly carrying these loads. All such points are collectively referred to as the "locus of rigidity" of the structure. For a bar structure, the locus of rigidity comprises all of the structure's joints, because loads applied at the joints can always be resolved into components that are oriented *along* the bars. For a sheet structure, the locus of rigidity comprises all of the structure's edges, where two sheets join at a dihedral angle. (The sheet material can be thought of as containing an infinite number of equivalent bars, connecting to every point along each edge.) Loads applied anywhere along any edge can be resolved into components that lie within the planes of the sheet material, and therefore lie along an equivalent bar.

7.14 ■ VIBRATION IN STRUCTURES

When a structure has been designed to have a rigid shape, vibration is rarely a problem. In a rigid structure, the precise position (in six degrees of freedom) of critical components attached to the locus of rigidity of the structure cannot be deflected except by corresponding *in-plane* deflections of the material of the structure. Vibrational energy, on the other hand, is predominantly concentrated in *transverse* modes, which are *perpendicular* to the plane of the material of the structure. Points on the structure's locus of rigidity are generally not affected by these vibrations. The areas of a structure that are susceptible to transverse vibrations are regions that are *away* from the locus of rigidity, such as the centers of large sheets.

Large sheets can also provide a path for acoustical coupling from external excitation sources. For this reason, in environments exposed to appreciable acoustical vibrations, it is a good idea to provide holes in large-sheet areas. Also, ribs can be added to large-sheet areas to increase their stiffness, thus reducing the amplitude of transverse vibrations.

7.15 ■ TIPS AND TECHNIQUES FOR CARDBOARD MODELING

Why Make a Cardboard Model?

After you have designed a rigid structure, it is a good idea to confirm that it is rigid by making a model. The purpose of a model is to investigate the rigidity of a structural design by scaling the forces and deflections of the structure to a level that is easily detected in your bare hands. Bar structures can be modeled with 1/8-in. diameter wooden sticks that connect at flexible joints. Joints made of hot-melt glue are rubbery enough to be flexible, yet still provide a stiff connection for axial forces. If your structure is to be made of sheet metal, vacuum formed, injection molded, or cast, cardboard is an ideal material to use in constructing your model. The short time it takes to make the model and the few minutes spent testing it will be time well spent. If you overlooked any of the constraints needed to make the structure rigid, it will be obvious when the completed model flexes in your hands.

If the structure is inadvertently overconstrained, that too will be obvious if you have to *force* the last joint together. Often, as your design takes shape in three-dimensional form, you will get new ideas that did not occur to you when the design was in your head or on paper. Also, the model will be a very efficient way of conveying your structural concept to others.

Types of Cardboard

Tablet back cardboard, typically about 25 to 30 mils thick, is a handy material for making models. It is abundant in 8½ × 11 inches and is also available in larger sizes. Drawing paper and diazo paper come sandwiched in this type of cardboard. One problem with this type of cardboard, however, is that it is of variable quality and it sometimes delaminates under the stresses induced near glue joints. If your model is to be "quick and dirty," you may not care. If you plan to work very hard on a "masterpiece," then be aware of the quality of your cardboard. Poster board is of more consistent quality and is available in a variety of thicknesses. Manila folders and hanging file folders are about 10 mils thick and are good modeling materials. There is another type of paper, called index stock, which is about 8 mils thick. This is about as thin as you will want to use. Models I have made with this have had overall dimensions in the 5 to 6 in. range. There are some advantages to using index stock over the thicker materials. For one, it cuts easily with scissors. Another advantage is that it is thin enough to go through a copier. This means you can lay out your part outlines, bend lines, and glue lines once on a sheet of paper, then copy enough "raw stock" to make several models. You might do this to investigate the relative stiffness of several similar models having slightly different configurations (more about this later). Corrugated cardboard and foam core material are not permitted for modeling rigid structures. The reason is that these materials both possess considerable flexural stiffness. A rigid structure, by definition, does not depend on the flexural stiffness of any of its members for its overall stiffness. All forces in a rigid structure are transmitted within the plane of the material in that structure. Therefore, the ideal material to use in a model, whose purpose is to investigate the rigidity of a configuration, is a material with *very low flexural stiffness*.

Gluing

The very best method for gluing models together is to use a hot-melt glue gun. It is fast and strong. The glue comes out of the heated nozzle of the glue gun with sufficient liquidity to "wet" the cardboard. It then cools (ordinary glues have to dry) to a hard, rubbery consistency in about 30 sec.

For gluing over an appreciable area of two surfaces, the hot-melt glue gun is not suitable because (1) it takes too much time to apply the glue to a large area, (2) the large area of contact further accelerates cooling, and (3) it is difficult to squeeze the glue to a uniform thickness so the joint turns out lumpy. For gluing large areas, ordinary white glue is good to use. The only trouble with it is that you have to wait for it to dry. If you use white glue to join the edge of one piece to the face of another, it is helpful to make a small flange to provide a contact area between the two pieces.

Size and Scale Considerations

The right size for your model is a size that fits comfortably in your hands. This can be interpreted to include models having overall dimensions ranging from about 5 in. to 15 in. Once you have decided on a scale for the overall dimensions of your model, you must decide what thickness cardboard to use. There is an advantage to making a model with material that is *thinner than scale*. If you compare two similarly sized models, one of which is rigid and one nonrigid, the difference in their stiffness will be *amplified* if you build them with thinner material.

On the other hand, by making scale models where the *thickness* of the material is also *to scale*, I have found that the measured stiffness improvement of a rigid steel frame with respect to a nonrigid steel frame is fairly well predicted by measuring the stiffness improvement of scale cardboard models of the same configurations. This can be an extremely powerful tool. You can predict whether a proposed structural modification will result in a 20% improvement in stiffness, double the stiffness, or produce an order of magnitude improvement, and you do not have to do any calculations or computer modeling. The stiffness measurements can be made by using a dial indicator and a spring scale. If, for example, you want to measure the twist stiffness of a structure that has a rectangular base, you could block up three corners and measure the deflection of the fourth corner caused by a vertical load applied at that corner, as described in section 7.11.

7.16 ■ STRUCTURALLY SIGNIFICANT POINTS

The job of the structure of a machine or instrument is to rigidly connect all *structurally significant points*. Structurally significant points are the points that must maintain a rigidly fixed dimensional relationship for any reason, whether subjected to significant forces or not. They include shaft bearing locations, actuating cylinder, solenoid and motor mounts, linkage pivot points, lens and mirror attachment points, "ground" locations for adjusting mechanisms, points at which the structure will attach to other objects or the floor, and so on.

All structurally significant points on a structure will be rigidly connected if the structure is three-dimensionally rigid and the structurally significant points all lie on the locus of rigidity.

7.17 ■ FORCE PATHS

Portions of some structures may be called on to carry relatively large loads, for example, frames for presses of various kinds. When designing such structures, it is desirable to keep major forces confined to short, direct, "single-member" paths within the structural material. This practice makes it easy to minimize small structural distortions that may occur during operation of the functional mechanism.

In the design of instruments, it is *especially* important to keep the force path short. A good instrument design will usually contain two separate structural paths, one for carrying forces and a second, separate structural path by which measurements are made.

7.18 ■ DESIGNING STRUCTURES FOR MACHINES AND INSTRUMENTS

The structure for a machine or instrument should be three-dimensionally rigid to avoid unwanted deflections and vibrations. Such a structure can be designed for any custom application by following a four-step process.

First: Identify all structurally significant points. It is the structure's job to rigidly connect all of these points.

Second: Design a *rigid core structure* that fits within the envelope of the proposed machine. The rigid core must be a three-dimensionally rigid structure. Because it is the purpose of the rigid core to provide three-dimensional rigidity to the entire machine structure, the volume of the rigid core should be appreciable compared to the overall volume of the machine. Preferably, it should be no less than 25%.

Third: Use the process of 3D joint addition to extend the locus of rigidity from the rigid core to all structurally significant points.

Fourth: Build a cardboard model and test it for rigidity.

7.19 ■ CONNECTING BETWEEN RIGID STRUCTURES

Often, within a structure, we can identify smaller structures which have been connected together. Visualizing structures this way can be a useful technique, both for analyzing existing structures, and creating new ones.

Two Rigid Structures with No Points in Common

Consider the structure needed to rigidly connect the two triangles of Fig. 7.19.1.

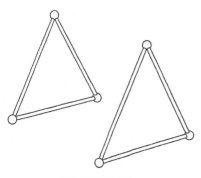

FIGURE 7.19.1

From our knowledge of 3D connections between objects, we know that six constraints and, therefore, six bars will be needed. One arrangement of bars that provides an exact constraint connection between the two bodies is shown in Fig. 7.19.2. This consists of three pairs of constraints, where the constraints of each pair are intersecting. This constraint pattern is similar to that of the tripod discussed at the beginning of this chapter.

Two Rigid Structures with One Common Point

When two rigid structures share a single common point, as in the case of the two tetrahedra shown in Fig. 7.19.3, only three degrees of freedom remain between the two bodies, namely, three **R**s intersecting at the common point.

To remove these three degrees of freedom, three constraints are needed. Three constraints (bars), positioned in a trilaterally symmetrical pattern and connecting from each of the remaining three apexes of one tetrahedron to those of the other tetrahedron will constrain these three **R**s provided the tetrahedra are not oriented so that these three constraints are parallel (or nearly parallel). The three constraints must be

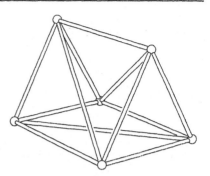

FIGURE 7.19.2

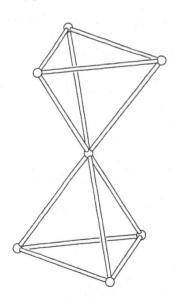

FIGURE 7.19.3

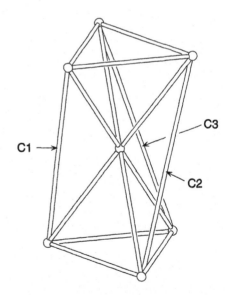

FIGURE 7.19.4

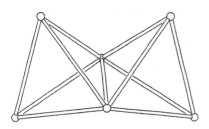

FIGURE 7.19.5

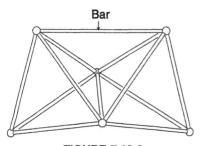

Bar

FIGURE 7.19.6

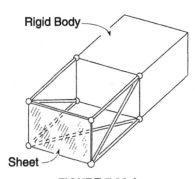

Rigid Body

Sheet

FIGURE 7.20.1

skewed. The first constraint, $C1$, removes one R, leaving two Rs which reside on a disk of radial lines in the plane defined by $C1$ and the common apex. $C2$, defining a second plane with the common apex, removes a second R, leaving only one R, which lies along the intersection of the first and second planes. The third constraint, $C3$, to be effective in removing the third R, must "exert a moment" about the third R. If the three constraints were parallel, $C3$ would be parallel to the last remaining R and unable to constrain it. Therefore, we conclude that the three constraints must be skewed, not parallel.

The demountable lens grinding tool shown in Fig. 6.1.7 used exactly the same pattern of constraints.

Two Rigid Structures with Two Common Points

As another example of how two smaller rigid structures can be connected to form a larger rigid structure, consider two tetrahedra that share a common edge, as shown in Fig. 7.19.5.

These two rigid bodies have one degree of freedom relative to each other, namely, rotation about their common edge. Therefore, we know that only one additional bar is needed, as shown in Fig. 7.19.6.

7.20 ■ CONNECTING BETWEEN FLEXIBLE STRUCTURES

Consider a rectangular sheet whose equivalent bar structure has a single diagonal. This structure, although two-dimensionally rigid, is flexible as a 3D structure. It has a single degree of freedom. Thought of as two (rigid) triangles sharing one common bar, the two triangles have a single R between them located along their common bar.

We can make a rigid connection between such a flexible structure and another (rigid) structure by using the usual six constraints *plus a seventh* constraint to offset the single degree of freedom that this structure has intrinsically. Fig. 7.20.1 shows a pattern of seven constraints being used to rigidly connect a sheet to a rigid body.

Rigid connections to underconstrained structures require an additional constraint in the connection for each degree of underconstraint in the structure.

How many constraints would be needed to connect between *two* structures, each flexible in one degree of freedom? Design the structure.

7.21 ■ CONNECTING TO POINTS NOT ON THE LOCUS OF RIGIDITY

It is not always convenient to make connections to structures on or near the locus of rigidity. Points on the locus of rigidity, by definition, are three-dimensionally rigid. But points throughout the area of a sheet are two-dimensionally rigid, and points all along bars are one-dimensionally rigid. So, if we need to make a connection to a structure that is required to provide constraint *only* in a direction that lies within the plane of a sheet, we can make our connections at points anywhere on the sheet.

Similarly, if we need to make a connection to a structure that is required to provide constraint *only* in a direction that is along the axis of a bar, then we can connect anywhere along that bar.

As an example of a rigid connection between two three-dimensionally rigid bodies, connected at points *off* the locus of rigidity, consider the connection between body A and body B in Fig. 7.21.1. Body A is a cubical sheet structure made of thin sheet material. The locus of rigidity for body A includes all of its edges. Points on the faces (away from the edges) are not on the locus of rigidity. These points can be easily deflected in a direction normal to the face. Nevertheless, we intend to make a rigid connection to body A by using just these points (in the centers of the faces). Body B, which is not the body of interest, is assumed to be three-dimensionally rigid with its locus of rigidity extending to the ends of the three posts that protrude from its top face.

Body A and body B are bolted together by three bolts through the center of each of three adjacent faces of body A. Each bolted connection provides two constraints through the center of the face and in the plane of the face. This pattern of three intersecting

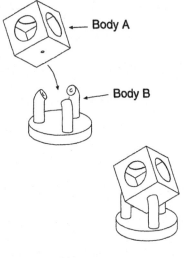

FIGURE 7.21.1

pairs of **C**s lying in three orthogonal planes provides a rigid connection to body A.

7.22 ■ WARP AND DISTORTION CAUSED BY WELDING

Welding can result in some unwanted distortion. As a result, there is a commonly held notion that welding is to be avoided in any part or structure where accuracy of shape is required. What is *not* commonly known is that *underconstrained* structures are much more severely plagued by weld distortion than are *three-dimensionally rigid* structures. To appreciate why this is so, we must first understand the mechanism that produces weld distortion.

Consider what will happen if a weld bead is made across one face of a rectangular metal bar. The molten metal in the bead attains a very high temperature. When the molten metal cools enough to resolidify, it is still hotter than the metal (which did not melt) on the other side of the bar opposite the bead. So far, the bar is straight.

As the bar equilibrates to a uniform temperature, the metal in the bead area will have cooled through a larger temperature difference than the metal on the other side of the bar opposite the bead. Therefore, due to the coefficient of linear thermal expansion, the bead side of the bar will shrink more than the other side. The bar will bend as shown in Fig. 7.22.2. Notice that the *length* of the bar does not appreciably change.

It is a characteristic of a three-dimensionally rigid structure to depend on the length of its component members (not on their straightness or shape) for its overall shape. Therefore, exactly constrained structures are relatively immune to welding distortion. A good example is a three-dimensionally rigid steel frame whose component parts are welded together in a welding fixture. After the final weld is made and the frame cools, it is removed from the fixture. Careful measurement reveals no discernible distortion. The features that were clamped in the welding fixture retain their exact relative positions when unclamped.

There are instances, however, where welding *can* influence the shape of a rigid structure. How can this be? From what has been said above, we know this can only happen when welding produces a change in length of a component member of the rigid structure. To see how this happens, let us look at an example.

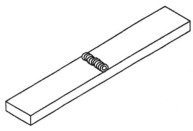

FIGURE 7.22.1

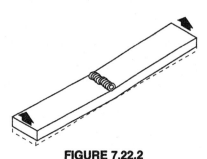

FIGURE 7.22.2

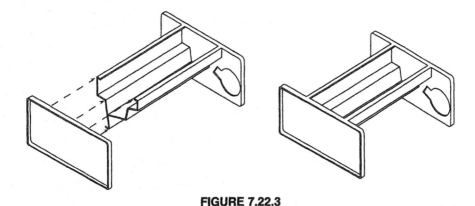

FIGURE 7.22.3

Figure 7.22.3 shows a frame that is made of 0.078-in. thick aluminum. We immediately recognize that this frame has a flexible shape. It consists of an open channel with end plates and, therefore, has one degree of freedom.

By adding a diagonal bar across the open face, we can supply the missing constraint and achieve a rigid structure. It so happens that it is a requirement for this frame for the four pad areas, A, B, C, and D, to be coplanar. To achieve this, the frame is placed in a welding fixture prior to welding the diagonal bar in place. The welding fixture positions pads A, B, C, and D so that they are coplanar. Now the diagonal bar is welded in place across the open side.

When the frame is removed from the fixture and measured, it is found that the pads A, B, C, and D are no longer coplanar. In fact, if we allow pads A, B, and C to determine a plane, pad D is measured to be high by about ⅛ in. This says that the diagonal measurement of the frame (the corner-to-corner distance measured along the diagonal bar) must have decreased. (The actual change in length can be calculated. See appendix.)

This is not all that surprising. When the diagonal bar is heated by the welding, it becomes hotter and therefore becomes longer. The diagonal measurement of the frame, however, remains the same because the frame is clamped in the welding fixture. Just as the final weld solidifies, the structure is in its correct shape and the diagonal bar is at an elevated temperature with respect to the rest of the structure. Later, when the frame is released from the fixture and has equilibrated to a uniform temperature, the diagonal

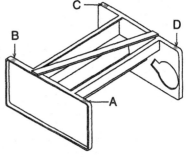

FIGURE 7.22.4

bar will have cooled a greater amount and, therefore, shrunk a greater amount than the rest of the structure. This results in the frame twist that we observe.

This effect is far more severe with aluminum than it is with steel for two reasons:

1. Aluminum has a higher coefficient of linear thermal expansion.

2. Aluminum's thermal diffusivity is much higher. This produces a much higher average temperature in the diagonal bar for a given amount of welding time.

Now with our understanding of how the distortion was produced, we can propose a remedy.

The solution to the problem is to devise a configuration for the missing constraint that does not allow cooling shrinkage to change the diagonal length. By replacing the diagonal bar with a "V" and welding the "base" of the "V" last, we solve the problem. The "V" still provides the needed constraint, but it is symmetrical. It does not allow shrinkage to shorten preferentially either diagonal distance relative to the other.

The "V" was welded to the frame while the frame was clamped in the fixture, with pads A, B, C, and D coplanar. When the rigid frame was unclamped, pads A, B, C, and D remained coplanar within our ability to measure them.

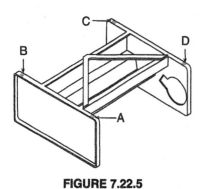

FIGURE 7.22.5

CHAPTER SUMMARY

In this chapter we used exact constraint design principles to design rigid, light-weight structures. The technique used was the same as that developed in Chapter 4 for flexures, recognizing the orders of magnitude ratio in stiffness between bending and stretching. By using this technique, a designer can quickly develop the ability to evaluate structural shape and determine the presence or lack of rigidity at a glance. This type of qualitative analysis is useful in the conceptual stage of design, and it provides a better intuitive understanding than one can achieve from studying the results of a finite element analysis.

APPENDIX

Calculating the Distortion of a Five-Sided Box Due to a Change in the Diagonal Length of the Open Face

Consider a five-sided box of dimensions $H \times W \times D$, which is open on the top. Imagine that the box is supported at corners 1, 2, and 3 so that those three points form a fixed plane. Because the box is underconstrained in one degree of freedom, we know that corner 4 will be free to move in a direction perpendicular to the plane defined by corners 1, 2, and 3. In fact, there is no reason that we should not think of the box as a mechanism consisting of two rigid bodies connected at a hinge.

Using what we know about the equivalence of sheet and bar structures, we can draw two tetrahedron bar structures (rigid by themselves) hinged along the line 1–3.

We can now describe the motion of body 1–3–4–8 relative to body 1–2–3–6, which we shall regard as fixed. For clarity, we define a set of orthogonal axes where Y is the hinge line and X goes through point 4.

The distance from the Y-axis to point 4 is defined as

$$a = \frac{D \times W}{l}$$

where l is the distance from point 1 to point 3 (the diagonal dimension of the base of the box.

Now, let us operate the mechanism. Suppose that point 4, initially in the X–Y plane, is deflected in the Z direction by a small amount, δ. Point 8 will move a corresponding amount δ in the Z direction and will also move an amount $H\delta/a$ in the negative X direction. We are interested in finding the change, Δl, in the distance between points 6 and 8, which corresponds with the deflection of point 4 by an amount δ. It can readily be seen (by similar triangles) that

$$\frac{\Delta l}{H\delta/a} = \frac{a}{l/2}$$

This simplifies to

$$\Delta l = \frac{2H}{l}\delta$$

or

$$\delta = \frac{l}{2H}\Delta l$$

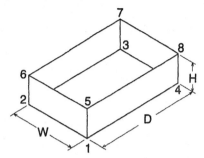

FIGURE 7.A1

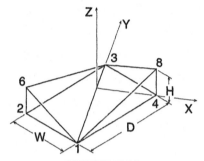

FIGURE 7.A2

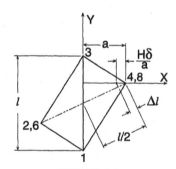

FIGURE 7.A3

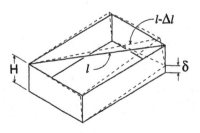

FIGURE 7.A4

CHAPTER 8

Exact Constraint Web Handling

A web is generally defined to be a strip of flexible material whose width is much greater than its thickness and whose length is much greater than its width. Photographic film and paper are two examples. Such webs can be easily damaged if mishandled by conveyance machinery. Therefore, web conveyance machinery needs to be carefully designed to avoid overconstraint in the connection to the web. By achieving an exact constraint *connection between the web and the web conveyance machinery, not only do we avoid damaging the web, but we can also achieve precision web tracking performance that would otherwise be impossible.*

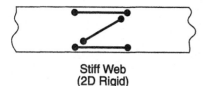

Stiff Web
(2D Rigid)

FIGURE 8.1.1

8.1 ■ WEB TYPES

Webs are divided into three basic types according to their two-dimensional stiffness characteristics.

Stiff webs, provided they are under tension, are two-dimensionally rigid according to our definition given in section 7.5. A stiff web will rigidly resist two-dimensional distortion. The bar equivalent structure of a stiff web contains edge bars *and* a diagonal bar. Photographic film and photographic paper are generally considered to behave as stiff webs, provided they are conveyed over spans that are not extremely long compared with their width. (The idea here is that if a span is very long compared with its width, the two edge bars and the diagonal bar would approach the condition of becoming so close together as to be practically coalesced into a single bar.) This chapter is primarily about stiff webs (except sec. 8.14).

With a *compliant web*, we must abandon the assumption that the web is two-dimensionally rigid. A compliant web might be one that is conveyed over a span length that is much greater than its width, or one that is subjected to high tension loads. Under these circumstances, the web can no longer be relied on to maintain its two-dimensional shape in response to loads applied by the conveyance apparatus.

Figure 8.1.2 shows the (exaggerated) response of a compliant web span to an off-center tension load. (This span had straight edges in its unloaded state.)

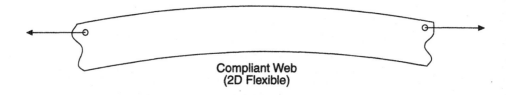

Compliant Web
(2D Flexible)

FIGURE 8.1.2

Crowned rollers can be used in the conveyance of compliant webs. The mechanism by which crowned rollers work is discussed in 8.14.

Fabric webs have an equivalent bar structure that contains the edge bars but *not* the diagonal bar. A fabric web does not resist two-dimensional distortion. A rectangular shape can be easily distorted into a parallelogram. We will not deal with the conveyance of fabric webs.

8.2 ■ A 2D PROBLEM

To begin, consider a stiff web supported on two rollers as in Fig. 8.2.1. Assume the web is under tension by some means not shown but that it is otherwise free of additional constraints. The span of web between the two rollers is just like the two-dimensionally rigid sheet of cardboard described in section 1.1. The cardboard is supported by the table. The span of web is supported by the rollers. Tension makes sure that the web does not sag in the middle. Like the sheet of cardboard, the web is considered to be a two-dimensionally rigid object. Also, like the sheet of cardboard, we are not interested in where the plane of the web is, only where the web is in its plane.

We care only about the web's position in X, Y, and θ_z. In other words, our problem is 2D. This makes things easy for us in a couple of ways. First, and most obviously, we only have to be concerned with three degrees of freedom instead of the usual six. Actually, it is even easier yet. Because the web is usually driven (in X) by one of the rollers, we rarely need to think about constraining this degree of freedom. It is just "understood." So now we have only two remaining degrees of freedom. The second benefit we get from reducing the problem to 2D is that we can make a single-plane diagrammatic representation of the web called a *web plane diagram*, even though in reality the web path is not confined to one plane. Figure 8.2.2 is such a diagram of the portion of web path depicted in Fig. 8.2.1. The "tails" of the web, which overhang beyond the rollers, are not in the same plane as the segment of web between the rollers. Nevertheless, we can diagram them as if they were. Although we acknowledge that the web is wrapped 90 degrees around each roller, we choose to diagram it as if it were unwrapped and flat. This will prove to be a handy technique for analyzing the constraints applied to a web and it further reinforces our premise that this is a 2D problem. Let us now examine the nature of the connection between each roller and the web.

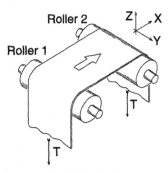

FIGURE 8.2.1

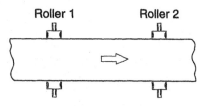

FIGURE 8.2.2

8.3 ■ EACH ROLLER IS A SINGLE CONSTRAINT

Because of the web's substantial angle of wrap around each roller (90 degrees in this case), web tension produces significant radial pressure of the web against the roller surface. This pressure, in turn, results in a considerable frictional coupling between the web and the roller surface. For purposes of web handling, this frictional coupling behaves as a *mechanical connection* between the web and roller.

Each roller thus imposes a constraint on the web in the direction parallel to the roller axis. The constraints are represented using the usual constraint symbol (●——●). Recall our definition of a constraint: Points on the object along the constraint line can move only at right angles to the constraint line, not along it. This is precisely the effect of the roller on the web.

An analogous connection is a pair of roller stock stands supporting a long piece of lumber. Each roller imposes a constraint on the board in a direction parallel to the roller axis.

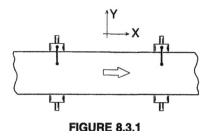

FIGURE 8.3.1

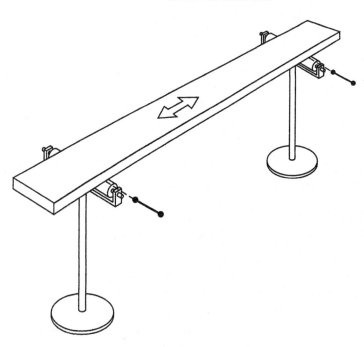

FIGURE 8.3.2

Another analogous connection is the one between the wheels of a vehicle and the road. The pavement moves easily past the vehicle in the direction the wheels are aimed but resists motion sideways. (Try pushing a car sideways.)

Referring now to Fig. 8.3.1, the effect of the two constraints on the two-dimensionally rigid web is exactly the same as the effect of two constraints on the sheet of cardboard in Chapter 1. The web is constrained to pivot about an instant center of rotation located at the intersection of the two constraints. In the case where the two rollers are parallel, the instant center will be located at infinity, thus permitting X translation of the web. But beware the first trap that snares the unwary designer. Although it looks like the web ought to track perfectly in X, being exactly constrained by two parallel rollers, we find that in practice the web will wind its way right off one end or the other of the rollers. After making this disappointing observation, a designer might conclude that the rollers are *not* properly aligned and then spend some considerable effort trying to improve that alignment.

As it turns out, such efforts are completely futile, and to see why, we need to look no further than to our analogy of the car. Imagine a driverless car whose steering wheel has been carefully aligned and fixed so the car will turn neither to the right nor left. The car is then placed on a long stretch of straight road and set in motion. If the car were carelessly aimed, it would clearly drive off the edge of the road after only a short distance. Now imagine how difficult it would be to achieve an alignment so good (of both the steering wheel and the initial starting angle of the car relative to the road) that the car would stay on the road indefinitely. It would not be possible. Thus, we have discovered the *main* difference between the constraint imposed by a wheel and that imposed by a link or a contact point, namely, that the constraint imposed by a wheel refers to no particular feature of the road. This is exactly the situation we have with our web on a roller. In this sense, the constraints imposed by wheels and rollers are different from those imposed by such devices as links or contact points.

The same problem faces us with the web and roller configuration of Fig. 8.3.1. Even if the rollers could be aligned perfectly and the web were perfectly straight, it would track off the rollers if its initial

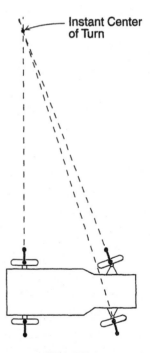

Instant Center of Turn

FIGURE 8.3.3

alignment were not also perfect. Acknowledging that such perfection is not possible, we look for another solution. We look first at how the problem is solved in a car. In a car, a servo (driver) monitors the car's cross-track position (Y) and makes appropriate steering corrections. When the car's steering wheel is turned, the axes of the front wheels are rotated so that the point at which they intersect the axis of the rear wheels shifts from infinity to some finite position, as shown in Fig. 8.3.3. This point defines the location of the car's instant center of rotation. The steering wheel thus controls the position of the car's instant center along the line of the car's rear axle. To steer the car, the driver must roughly match the position of the car's instant center with the location of the road's center of curvature and also make smaller corrections as needed. Clearly, we could apply the same solution to keep our web on the rollers.

To accomplish this with our web, we might install a sensor to monitor the web's cross-track position then use the information from this sensor to make adjustments to the steering angle of one of our rollers, say roller 2. For example, say the web sensor finds the web has moved too far in the positive Y direction. This would signal a change in the steering angle of roller 2, which would in turn cause the web's instant center to shift to the position shown in Fig. 8.3.4. The web would then rotate about this instant center, causing the web edge to move in the negative Y direction. Thus, the cross-track motion of the web can be controlled.

The point of all this is not to investigate the design of a servo. It is to illustrate that the rollers impose very real constraints on the web (albeit without regard to the web edge) and the web responds *exactly* as we would expect it to. When we apply two constraints to the web, this will define an instant center about which the web will be constrained to rotate. This knowledge enables us to control the web and understand its behavior, whether we choose to control the web's cross-track position with a servo or by some other means.

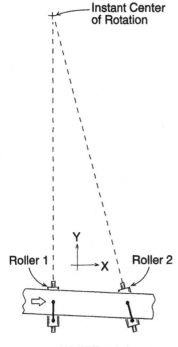

FIGURE 8.3.4

8.4 ■ EDGE GUIDES

Another means of controlling the web's cross-track position is to replace the upstream roller with an edge guide. An edge guide, such as the one illustrated in Fig. 8.4.1, may consist of nothing more than a pivoting paddle against which the web edge bears (under the influence of a nesting force generated by some means, not shown). An edge guide imposes a constraint on the web *at right angles to the web edge*. This differs from the constraint applied by a roller in just two ways. First, the constraint applied by a roller does not reference any particular feature of the web, whereas the edge guide references the web edge. The second difference, more subtle, is that the position and angle of a constraint applied by a roller remain fixed (with respect to the roller axis), whereas the angle of the constraint applied by an edge guide shifts as the angle of the web edge shifts, always remaining square to the web edge. This may sound insignificant, but it is not, as we shall soon see.

Figure 8.4.2 is a schematic of a web constrained by an edge guide and a roller, spaced apart by a distance l. The web is shown misaligned considerably. The two constraints define an instant center of rotation for the web at their intersection. As the roller drives the web in the direction indicated, the web responds by rotating clockwise about its instant center.

We are interested in observing the *cross-track* position of the web as it is conveyed in the direction shown. Specifically, we will examine the y position of the web edge as it crosses the roller axis. At time $t = 0$, $y = y_0$. If we look at point A on the web edge, it is in contact with the roller and positioned directly over the roller axis.

After a short interval of time, Δt, point A will move Δx (perpendicular to the roller axis) to point A'. As a result, the web edge (at the roller) will have moved $-\Delta y$. Thus the web velocity has two components: an *in-track* component $v_x = \frac{\Delta x}{\Delta t}$ and a *cross-track* component $v_y = \frac{\Delta y}{\Delta t}$.

For small intervals of time:

$$\frac{-\Delta y}{\Delta x} = \frac{y}{l}$$

stated in terms of velocity:

$$\frac{v_y}{v_x} = \frac{-y}{l}$$

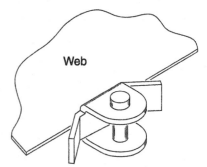

FIGURE 8.4.1

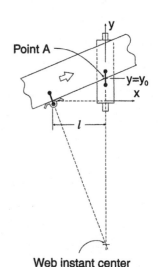

FIGURE 8.4.2

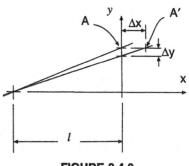

FIGURE 8.4.3

$$-\frac{v_x}{l} = \frac{v_y}{y} = \frac{1}{y}\frac{dy}{dt} = \frac{d}{dt}(\ln y)$$

$$d(\ln y) = \left(\frac{-v_x}{l}\right)dt$$

Integrate both sides (l and v_x are both constant):

$$\int d(\ln y) = \ln y = \int \frac{-v_x}{l}dt = \frac{-v_x t}{l} + c_1 = \frac{-x}{l} + c_1$$

$$y = e^{\left(-\frac{x}{l} + c_1\right)}$$

plug in initial conditions: when $x = 0$, $y = y_0$

$$y_0 = e^{0 + c_1} \rightarrow c_1 = \ln y_0$$

substitute value of c_1:

$$y = e^{\left(\ln y_0 - \frac{x}{l}\right)} = e^{\ln y_0}e^{\frac{-x}{l}} = y_0 e^{\frac{-x}{l}}$$

Thus we see the that lateral position of the web edge at the roller is a decaying exponential function of the *distance* the web moves in track. In engineering, we are accustomed to the notion of a decaying exponential *time constant*—here we have a decaying exponential *length constant* equal to the distance l between the upstream edge guide and the downstream roller.

For a web span constrained by an upstream edge guide and a downstream roller and having a starting position displaced by an amount y_0 laterally from its steady-state lateral position on the roller, the lateral position of the web edge at the roller will follow a decaying exponential curve:

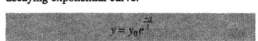

$$y = y_0 e^{\frac{-x}{l}}$$

where x is the *in-track* displacement of the web.

The web of Fig. 8.4.2 eventually reaches the steady-state position shown in Fig. 8.4.4, where the

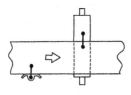

FIGURE 8.4.4

two **C**s applied to the web are parallel, thus defining an instant center of rotation (**R**) at infinity. This **R** is equivalent to a pure X translation (**T**). This is a very good way of achieving stable web conveyance while tracking the web's edge. However, one important drawback of this configuration is that it is *unstable* when operated in *reverse*.

8.5 ■ CAMBERED WEB

Up until now, we have considered the behavior only of webs having straight edges. Now, we will discuss the behavior of a cambered web. A cambered web is one whose edges are cut having a finite radius of curvature. We can understand the behavior of such a web by applying the same 2D analysis techniques we have used already.

Figure 8.5.1 shows a web with greatly exaggerated camber. This web is constrained by an edge guide positioned upstream of a roller. By construction, the locations of the center of curvature and the instant center of rotation have been found and are shown. As the web is advanced in the direction indicated, it must rotate about its instant center. This results in an apparent cross-track motion of the web edge at the roller in the negative Y direction. It also results in the center of curvature moving toward the constraint line of the roller. This, in turn, produces a new location for the web's instant center. The instant center thus moves along the roller constraint line *toward* the web's center of curvature.

As the web is further advanced, the center of curvature continues to move toward the roller constraint line and the web's instant center continues to move in along the roller constraint line toward the center of curvature. Asymptotically, the two points approach the condition shown in Fig. 8.5.2 where they are superimposed.

Recall what we learned about the steady-state position of a straight web constrained by an edge guide located upstream of a roller. The web tracked toward a stable position where the two constraints (from edge guide and the roller) were parallel (intersecting at infinity). Now, it so happens that the straight web's center of curvature is also at infinity. This is not a coincidence. We have discovered the following:

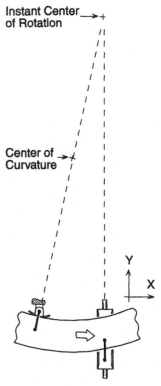

FIGURE 8.5.1

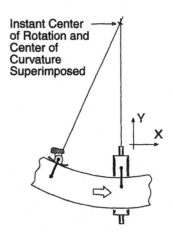

FIGURE 8.5.2

A web of any camber radius (finite or infinite), when constrained by an edge guide located upstream of a roller, will track toward the steady-state position where the web's instant center of rotation is superimposed with the web's center of curvature.

Furthermore, if we look closely at Figs. 8.4.4 and 8.5.2, we see that in this stable position, as the web crosses the roller, its edge makes a 90-degree angle with the axis of the roller. Regardless of the camber of its edge, a web constrained by an edge guide located upstream of a roller will approach the stable condition where the web edge crosses the roller square to its axis. For just this reason, we can think of such a downstream roller as an *angular constraint*. The roller *seems to constrain* the web edge to be at a 90-degree *angle* to the roller axis. But, as we have seen, this behavior only occurs when the upstream lateral position of the web has been fixed (as by an edge guide). Schematically, we can represent this stable constraint pair on our web plane diagram, as shown in Fig. 8.5.3. The edge guide is diagrammed as an arrow fixing the lateral position of the web edge at its head. The roller (angular constraint) is diagrammed as a line representing the roller's axis. The web edge is shown being constrained square to it. This combination of upstream lateral constraint and downstream angular constraint is frequently encountered in the design of web paths.

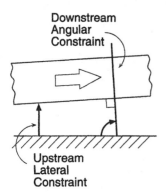

Downstream Angular Constraint

Upstream Lateral Constraint

FIGURE 8.5.3

8.6 ■ NECESSARY "PIVOTING" OF THE WEB ON THE ROLLER SURFACE

In the stable position shown in Fig. 8.5.2, where a cambered web rotates about its center of curvature, a slight θ_z "pivoting" motion is forced to occur at the interface with the roller, whose cylindrical surface is moving purely in X (in our web plane diagram). Near the web edge closest to the instant center of rotation, the web velocity is slightly *less* than the roller surface velocity. Near the edge farthest from the instant center, the web velocity is slightly *greater* than the roller surface velocity. Thus, a slight pivoting motion must occur about the approximate center of the area of contact of the web with the roller. This pivoting motion is analogous to that of the front wheels of a car relative to the road when the driver turns the steering wheel. When the car is parked, considerable effort is required to turn the steering wheel, as gross slip must occur at the tire–road interface. However, if the car is *moving*, only a small effort is needed. Under this condition, elastic creep of the rubber tires permits θ_z rotation with the application of only a small amount of torque. Similarly, elastic creep will work to our benefit at the web–roller interface. Sometimes, the web itself will be elastic enough to provide this "creep mode" behavior. If the roller has a rubber cover, then the rubber may be the dominant contributor to elastic creep.

This pivoting motion of the web on the roller surface occurs whenever the instant center of rotation of the web is not coincident with that of the roller surface in the web plane diagram. As we have just seen, this occurs when a cambered web runs on a cylindrical roller. It should be easy to imagine how pivoting would also occur if a straight web were running on a slightly conical roller. Another time this pivoting motion occurs is during the first several "length constants" of operation of a straight web initially misaligned in a web path comprising an upstream lateral constraint and a downstream angular constraint.

Forces Generated by Pivoting

The pivoting motion of the web relative to the roller, as we have seen, occurs whenever the instant center of rotation of the web is not coincident with that of the roller surface. Whenever this pivoting motion occurs, a balanced set of forces comes into play.

For example, consider the cambered web (camber shown exaggerated) shown diagramatically in Fig. 8.6.1 in its steady-state position as it is conveyed by an edge guide and a downstream roller. As a result of the pivoting motion that must occur at the roller–web interface, a clockwise torque Γ is applied to the web by the roller. To balance this torque, a force couple is generated, consisting of a force P1 at the edge guide and P2 at the roller. Although these forces are not a factor in determining the web's steady–state position (the web's steady–state position is determined by the location of its instant center), it is helpful to have an understanding of them. For example, the force that develops at the edge guide would obviate the need for a spring for this purpose.

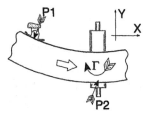

FIGURE 8.6.1

In operation, the web edge will just naturally ride against the edge guide.

8.7 ■ CONNECTION BETWEEN WEB AND ROLLER: "PINNED" VERSUS "CLAMPED"

A good conceptual model of the connection between the web and roller is to imagine that a single thumbtack has been used to secure the web to the roller surface at the approximate center of the contact area between the web and roller. Of course, this is only a conceptual model. The thumbtack would have to be repeatedly moved as the web advances. In this model, the connection between the web and roller consists of two constraints: an X constraint and a Y constraint. The connection is free in θ_z, as it needs to be. The Y constraint is the single, axially oriented constraint that we described in section 8.3. The X constraint, which has been tacitly ignored up until now, couples the web to the roller in the drive direction. Generally, one roller in a web path will be the *drive* roller. The X constraint between the drive roller and the web allows the drive roller to drive the web in the X (in-track) direction.

All other rollers in a web path are generally *idler* rollers. The X constraint between the web and each idler roller simply constrains the rotational position of each idler, which would otherwise spin freely.

We refer to this single–thumbtack model as a pinned connection between the web and roller. In general, when we design a machine whose purpose is to convey a web, the rollers of that machine are intended to behave according to the pinned model. We generally intend for each roller to have such a pinned connection to the web.

Occasionally, we intend for the connection between a web and roller to be *clamped*. In a clamped connection, θ_z pivoting of the web is not possible. A platen roller such as that of a typewriter or a pen plotter is a good example of a roller with a clamped connection. In both of these cases, the paper is effectively "attached" to the roller along the full length of the nip formed by a pressure roller. If the pressure roller were relatively short in length and centrally located relative to the platen roller, the "pinned" model might be appropriate. But with a long, aggressively loaded pressure roller, such as is used on typewriters and pen plotters, the connection between the paper and platen roller must certainly be considered to be a clamped connection.

A clamped connection between a roller and the web makes it impossible to change the *steering angle* of the roller relative to the web. Such a clamped connection completely constrains the web in 2D. Any additional constraints would be an overconstraint. Thus, we avoid using the clamped connection mode when we design web conveyance machinery. Almost always, we use the pinned connection mode.

8.8 ■ FLANGED ROLLERS

The flanged roller is an *overconstraint*. The roller itself is one constraint. The flanges, intended to act as an edge guide, are two more constraints. The constraints are superimposed, resulting in an *overconstraint*. The only way a flanged roller can succeed is if web tension and/or wrap angle are low enough to permit the web to slip on the roller surface (analogous to pushing the parked car sideways). Otherwise, the web edge will either be crushed by the flange or the web will climb over the flange. Both situations are ugly. If you really want to have an edge guide at a particular spot in your web path, then the constraint imposed by the roller has to go.

FIGURE 8.8

8.9 ■ ZERO–CONSTRAINT WEB
SUPPORTS

In the design of web conveyance apparatus, it is all too easy to end up with a web that is overconstrained. The web path layout will usually show the web being conveyed over many rollers. If each roller applies a constraint to the web, it is easy to see how the web would be grossly overconstrained. To avoid this problem, it is helpful to know about some special "rollers" that do not constrain the web. Collectively, we refer to these as *zero-constraint* web supports.

Nonrotating Shoes

One easy way to eliminate the constraint imposed by a roller is to lock the roller so that it cannot turn. Thus the roller becomes a fixed "shoe." By doing this, gross slip is guaranteed between the web and the surface of the shoe. The roller constraint is eliminated. Unfortunately, it is sometimes undesirable to

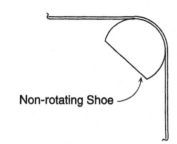

FIGURE 8.9.1

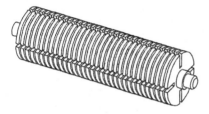

FIGURE 8.9.2

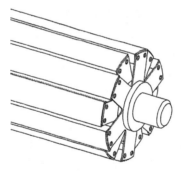

FIGURE 8.9.3

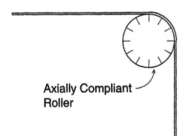

Axially Compliant
Roller

FIGURE 8.9.4

drag the surface of the web over a fixed surface because of scratching or the increased tension caused by the added drag. A way to overcome these problems is by pumping air into the web–shoe interface, forming an air cushion for the web to ride on. In general, however, this approach is expensive and has only found application in fixed manufacturing facilities.

The diagram symbol for a nonrotating shoe is shown in Fig. 8.9.1. The flat on this symbol is suggestive of the fact that it is nonrotating.

Axially Compliant Rollers

It is possible to eliminate the roller *constraint* without eliminating the roller by using a specially constructed roller whose surface has been designed to be *axially compliant*. The idea of an axially compliant roller is that the roller surface will freely move in the cross-track direction as demanded by the web and will then restore itself to its initial position after the web has left contact.

Here are two examples of axially compliant rollers. Figure 8.9.2 shows a roller with deep grooves cut in its thick rubber cover (US Patent #4,221,480). The roller surface could be described as being composed of a series of thin rubber disks. In operation, the disks will collectively deflect a small amount laterally as required, but will remain relatively stiff in the radial and tangential directions.

Figure 8.9.3 shows an axially compliant roller of a somewhat different construction (US Patent #5,244,138). This roller supports a web on 12 individual *slats*, or segments, each of which is connected at its ends by a sheet flexure to a rigid center shaft. Each segment is thus independently free to move axially a small amount in response to any misalignment between the web and roller. The segments can be undercut in the center section so that they support the web only near the edges. Flanges can be installed to provide lateral constraint between the web and roller. This can work two ways: either the roller can position the web laterally, or the web can position the roller laterally.

Figure 8.9.4 shows the diagram symbol for an axially compliant roller. This symbol suggests that the roller surface consists of a number of axially oriented slats that are individually free to move axially.

Castered Rollers

The castered roller is an excellent way to have the benefit of a rolling support for a web without imposing any constraint on the web. Actually, the constraint that exists between the roller and web does not go away. We simply add a degree of freedom (remove a constraint) to the connection between the roller and machine frame, thus allowing the web to position the roller rather than to have the roller impose its constraint on the web.

A castered roller self-aligns to the passing web in exactly the same manner as a castered wheel on a chair or cart self-aligns to the passing floor. A castered wheel has two cascaded rotational degrees of freedom: R_1, which is the wheel's own axis, and R_2, which is the yoke's pivot axis.

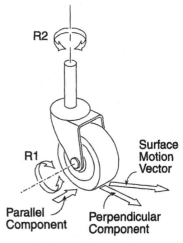

FIGURE 8.9.5

The motion of the surface in contact with the wheel comprises a component parallel to the wheel's axis and a component perpendicular to its axis. The perpendicular component produces rotation of the wheel about R_1. The parallel component has no alternative but to produce rotation of the yoke about R_2. This rotation about R_2 is in a direction that results in a gradual reduction of the misalignment of the wheel. Like the response of a web conveyed on an upstream lateral constraint and a downstream roller, the magnitude of the wheel's misalignment is a decaying exponential function of the caster radius (the distance between R_1 and R_2).

I have come across a simple (but incorrect) explanation for the behavior of a castered roller that makes pretty good intuitive sense but, unfortunately, is absolutely wrong. It goes like this:

The center of the contact area between the wheel and the surface with which it is in contact will always try to trail directly *behind* the spot where the caster axis intersects the surface. The center of the contact area and the caster axis will line up along the direction of motion just like a chain link will align with a tension force applied to the chain. If the roller is initially misaligned, such that the center of contact is off to the side, it is like having "slack" in the chain.

Although this explanation sounds plausible, consider what it would predict for the apparatus of Fig. 8.9.6, which depicts a castered wheel in contact with a surface that is moving in the positive X direction.

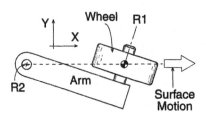

FIGURE 8.9.6

It predicts that the wheel would go to the position shown in Fig. 8.9.6, where all of the "slack" is taken up. In reality, however, when the caster is in this position, the web motion vector not only causes the wheel to spin about its axis, $\mathbf{R}_1$, but also has a residual component parallel to $\mathbf{R}_1$.

As noted, this will produce a counterclockwise rotation about $\mathbf{R}_2$, causing the assembly to gradually attain the position shown in Fig. 8.9.7. This is the position where the roller axis intersects the instant center of the moving surface.

It turns out that the axis, $\mathbf{R}_1$, of the castered wheel or roller will move toward the position where it intersects the instant center of rotation of the surface with which it is in contact.

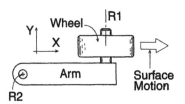

FIGURE 8.9.7

Suppose we have an exactly constrained web segment such as the one shown in Fig. 8.9.8. Notice that no constraints have been indicated, just an instant center of rotation. Obviously, two constraints must have been applied to the web to define this instant center. But we need not be concerned with the constraints. We only need to know that the web is exactly constrained and we need to know the location of its instant center.

If we want to support the web with another roller, that roller should be a zero constraint. If we opt to use a castered roller, as in Fig. 8.9.9, the roller will then automatically align itself so that its axis intersects the web's instant center of rotation.

Instant → •
Center
of Rotation
of Web

FIGURE 8.9.8

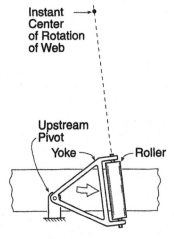

FIGURE 8.9.9

Let us now examine the *edge view* of the castered roller shown in Fig. 8.9.9. (Fig. 8.9.9 is a *web plane* view.) The apparatus might look as shown in Fig. 8.9.10. Rollers A and C are shown. Together, they have exactly constrained the web and determined the location of the web's instant center of rotation. To avoid overconstraint, roller B is castered about an upstream axis. The location of the caster axis is shown parallel to the angular bisector of roller B's incoming and outgoing spans and upstream of roller B by a distance that is roughly equal to the width of the web. This is the correct position for the caster axis regardless of the amount of wrap of the web around the castered roller.

In every case, the caster axis is positioned parallel to the bisector of the incoming and outgoing spans.

As a rough rule of thumb, the caster radius (the distance between the caster axis and the roller axis) should be about the same order of magnitude as the length of the roller (or width of the web). The exact value of the caster radius is not very critical.

As the web is advanced, a castered roller will approach the steady-state condition (where the roller axis intersects the web's center of rotation) as a decaying exponential function of the web's in-track displacement. After the web has advanced a distance of roughly five times the caster radius, the castered roller can be considered to have reached its steady-state position. Because the castered roller has a decaying exponential response when operated in the forward direction, it is unstable when operated in reverse. Note the similarity between the response of the castered roller with that of the web span constrained by an upstream lateral/downstream angular constraint pair (see section 8.4).

Web Plane Diagram Symbol for a Zero Constraint

In the web plane diagram, a zero-constraint web support (whether it is a fixed shoe, an air bearing, an axially compliant roller, or a castered roller) is diagrammed as a dashed line. Unlike the angular constraint symbol (which is shown as a solid line at right angles to the web edge), there is no angular relation between the web edge and the dashed line symbol for the zero constraint.

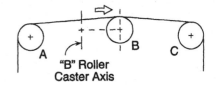

FIGURE 8.9.10

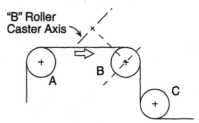

FIGURE 8.9.11

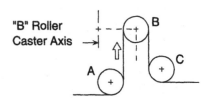

FIGURE 8.9.12

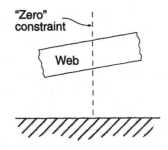

FIGURE 8.9.13

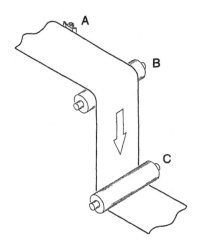

FIGURE 8.10.1

8.10 ■ GIMBALLING TO ACHIEVE A "JOINT" IN THE WEB PLANE DIAGRAM

Back in section 8.2, a couple of advantages were cited that owed to the fact that web-handling problems are two dimensional. There is one more big advantage. We can use the web's flexibility in twist to give us an extra degree of freedom. We can take advantage of one span's flexibility in twist to provide an adjacent span with an added degree of freedom. Consider the web path shown in Fig. 8.10.1, consisting of the combination of an upstream lateral constraint (A), a downstream angular constraint (B), and an additional roller (C).

The web plane diagram schematic of Fig. 8.10.2, which shows roller C with exaggerated misalignment, reveals that the web has been overconstrained.

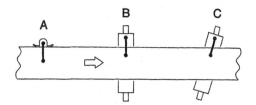

FIGURE 8.10.2

Constraint pair A–B defines one instant center, pair B–C defines a second, and pair A–C defines a third. The fragile web is being asked to rotate about three different instant centers at once. It is tempting to try to resolve this problem by setting the alignment of roller C so that its axis intersects the A–B instant center in the web plane diagram. This solution is satisfactory only if the web's camber radius never changes. However, when we try to track a web with a different camber radius (perhaps just a different section of the same web), the location of the A–B instant center will automatically move to the new location of the web's center of curvature. The C constraint will then no longer intersect the A–B instant center and we will again have a misaligned C roller.

One solution to this problem is to leave the C roller axis fixed and gimbal the B roller, as in Fig. 8.10.3. Gimballing the B roller means allowing it to pivot freely about an axis parallel to its incoming span. In this way, the web in span A–B will respond only to the constraints imposed by A and B.

Constraint C will have no influence on the web in span A–B other than to determine at what angle roller B is gimballed and, therefore, how much span A–B is *twisted*. But, as we pointed out in the beginning, we are not concerned with constraints that control *where* the plane of the web is, only with those controlling where the web is in its plane. So, in effect, by gimballing roller B we have added a degree of freedom. We can now consider the behavior of span B–C as if the web is laterally constrained at B (its position is determined by constraint pair A–B) and angularly constrained downstream at C. Figure 8.10.4 is the web plane diagram representation of this web path configuration. The gimballing of roller B is represented as a "joint" in the web plane diagram (WPD).

It is the *orthogonality* between the incoming and outgoing web spans that allows gimballing to produce a joint in the WPD. A 90-degree wrap works best. Wraps of 0 degrees and 180 degrees do not work at all. Therefore, as a rule of thumb, it is recommend-

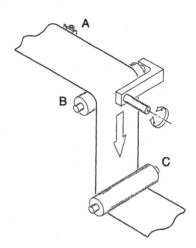

FIGURE 8.10.3

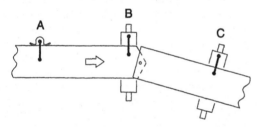

FIGURE 8.10.4

ed that to effectively achieve a joint in the WPD, the angle of wrap of the web around a gimballed roller should be 90 ± 45 degrees.

It should be emphasized that by gimballing the B roller in the manner described above, the single constraint applied by the B roller to the web is applied to the span A–B. In the manner described in section 8.3, the A and B constraints combine to define an instant center about which span A–B rotates. This determines the resulting lateral position of the web at B. Web span B–C then behaves as if there is an *implied* upstream lateral constraint at B, with roller C being the associated downstream angular constraint.

From this, one can easily imagine that the C roller could be gimballed, allowing a D roller to be added without overconstraint. By gimballing the D roller, an E roller could then be added, and so on. One could imagine a long web path consisting of a concatenated series of such spans from gimballed roller to gimballed roller, where each gimbal provides the degree of freedom needed by the following span and each roller acts as an angular constraint for the web in the preceding span.

8.11 ■ OVERHANGS

Another solution to the problem posed in the previous section is to leave the axis of roller B fixed and allow roller C to become a zero constraint (either by castering or making the surface axially compliant or making it a nonrotating shoe). Then, if we wish, we may gimbal C and add roller D. Figure 8.11.1 shows such an arrangement schematically. In this web plane diagram, schematic symbols have been used. The arrow symbol at A represents the edge guide. The solid lines at B and D represent those roller axes and are indicated to be perpendicular to the web edge. The broken line at C indicates that it is a zero constraint. In the sense that span A–C resembles a 2D beam supported by constraints at A and B, portion B–C is referred to as the *overhang*, because it is not between A and B, but rather *cantilevered* off the end.

If, for example, we decided to use a castered and gimballed roller at C, we must keep in mind that we are asking the web to exert a lateral force on the roller in order to get it to rotate about its caster axis. Furthermore, it is the extreme end of the web's overhang that will apply this lateral force to the web, so we must be careful to ensure that the lateral force needed at the end of the overhang does not exceed the maximum force that the web is capable of exerting.

The maximum force that the web can exert without buckling can be found by imagining the two diagonal lines sketched in Fig. 8.11.2 as two cables supporting tension T. It can then clearly be seen that one of the cables will just begin to go slack when the magnitude of F reaches the value:

$$F = \frac{1}{2}\frac{W}{l}\,T$$

In addition to the lateral force imposed on the overhang by the C roller, web tension in the C–D span also exerts a lateral component of force on the overhang. Depending on the magnitude of the mis-

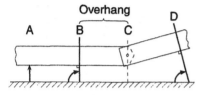

Overhang

FIGURE 8.11.1

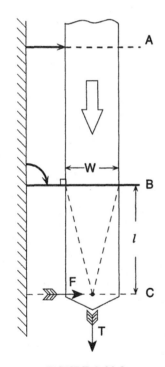

FIGURE 8.11.2

alignment of the C–D span, this could be a significant factor. Therefore, whenever an overhang is designed into a web path, care should be taken to ensure that applied forces do not cause the web in the overhang to buckle.

8.12 ■ EXAMPLE: WPD ANALYSIS OF A WEB CONVEYANCE APPARATUS

We are now ready to use the web plane diagram (WPD) to analyze a multiroller web conveyance apparatus. This will tell us the constraint condition that exists between the web and the conveyance apparatus. If the connection to the web is either *undercon-strained* or *overconstrained*, we can then make the necessary alterations to achieve an *exact constraint* connection. The web, when exactly constrained by the apparatus, will then be conveyed precisely and without damage.

For our example, we will analyze the web conveyance apparatus of Fig. 8.12.1, whose job it is to convey a thin plastic web loop in a clockwise direction around the perimeter of the apparatus. The web is supported on four rollers, labeled A, B, C, and D.

For the analysis, we start by making an edge view diagram (Fig. 8.12.2) to show any lateral constraints or zero constraints and to show the location of any caster or gimbal axes of each roller. Roller A, at the top of the apparatus, is gimballed. In addition, there is an edge guide provided at A. To avoid overconstraint, roller A has been made axially compliant. After roller A, the web travels to roller B, whose position is fixed. A motor, coupled to the shaft of roller B, provides drive for the web. Next, the web is conveyed to roller C, which is a castered and gimballed tension roller. After leaving roller C, the web is conveyed to roller D, which is an idler whose position is fixed. Finally, the web returns to roller A, completing the loop.

Now we are ready to begin the WPD. I like to start drawing the WPD on an edge guide, because the lateral position of the web is fixed and known. I then look downstream to find an associated downstream angular constraint. In our example, there is an edge guide at roller A. Roller B is the associated downstream angular constraint. The WPD can be started as in Fig. 8.12.3. The web is shown with its lateral position fixed at A (arrow symbol), and with its edge square to the axis of roller B. Owing to tolerances, the

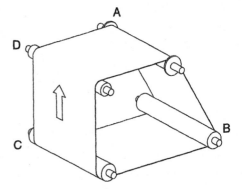

FIGURE 8.12.1

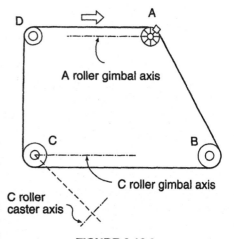

FIGURE 8.12.2

roller B axis will not be perfectly square to the machine frame. This is deliberately shown exaggerated in the WPD. By exaggerating the nonsquare angle between the B roller and the machine frame, the need for gimbal joints becomes obvious.

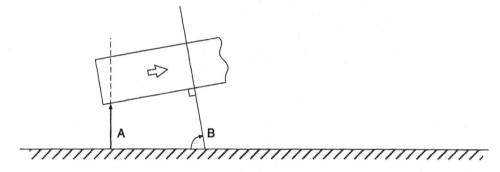

FIGURE 8.12.3

The web overhangs beyond roller B to roller C. Because the web is exactly constrained by A and B, no constraint is needed at C. Therefore C needs to be a zero constraint support. (Because C is castered, it *is* a zero constraint roller.) The zero constraint at C is indicated on the WPD by a dashed line representation of its axis.

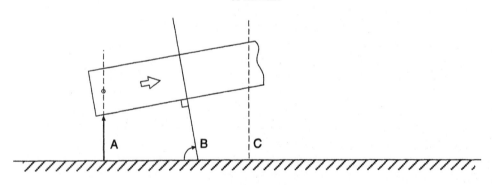

FIGURE 8.12.4

Because the web path of our example is a loop, we can anticipate the need to "close the loop" in our WPD. Figure 8.12.5 reminds us that in order to connect from C to D and back to A again, we are going to be need a couple of joints in the WPD. Those joints are provided by gimbals at A and C.

Indeed, if it were not for roller D, we could complete our WPD and it would look like Fig. 8.12.6.

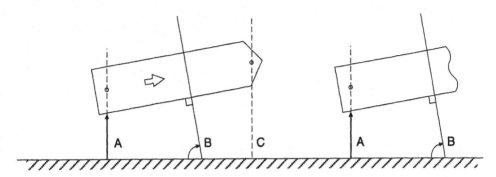

FIGURE 8.12.5

Axial compliance at A and caster at C provide the
necessary zero constraints at those rollers. Gimbals at
A and C provide the needed joints in the WPD.

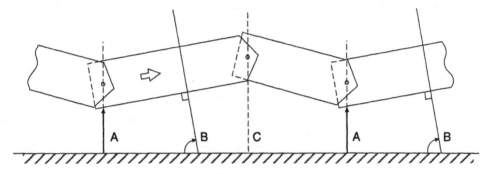

FIGURE 8.12.6

We must account for roller D, however. Roller D
tries to behave as a downstream angular constraint
relative to the fixed lateral position of the web at
roller C. The result is the WPD shown in Fig. 8.12.7.

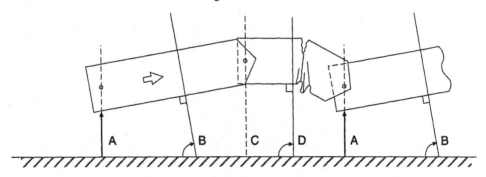

FIGURE 8.12.7

Roller D contributes an unwanted constraint to the web path. If we must have a roller at D, then it ought to be rendered a zero constraint by some means (fixed shoe, air bearing, axially compliant, or castered) or it should be gimballed. If we render the D roller a zero constraint, the web is conveyed from A to C as if it were not even there.

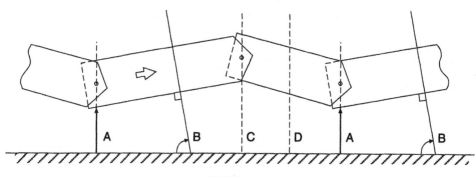

FIGURE 8.12.8

If, on the other hand, we choose to *gimbal* the D roller, the resulting WPD will be as shown in Fig. 8.12.9. The D roller will be the downstream angular constraint for the span leaving roller C. The web loop will then be closed by the short span from D to A. Gimbals at D and A will permit this. However, the WPD warns the designer that significant roller misalignments can result in steep gimbal angles at D and A due to the short span between D and A.

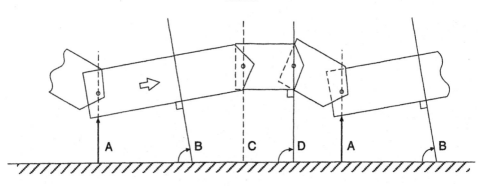

FIGURE 8.12.9

8.13 ■ TWO-ROLLER BELT TRACKING

The situation where a stiff web belt is supported on just two rollers represents a special and interesting case because it poses a couple of difficulties that are either not encountered or are easily avoided in systems having three or more supports:

- In a two-roller belt configuration, it is impossible to achieve the stable configuration of an upstream lateral constraint followed by a downstream angular constraint without simultaneously having the unstable combination of an upstream angular and a downstream lateral constraint. Because there are only two rollers, each roller is simultaneously upstream and downstream of the other.

- The practice of gimballing a roller to achieve a "joint" in the WPD does not work with 180-degrees of wrap. Gimballing to achieve a "joint" in the WPD works by taking advantage of one span's flexibility in twist to provide the other span a needed degree of freedom. This is done optimally at 90 degrees of wrap. It does not work with 180 degrees of wrap. Unfortunately, when we have only two rollers (of roughly equal size) both rollers have about 180 degrees of wrap.

Because of these two difficulties, the usual process of designing or analyzing a web path using the WPD is not especially useful in the two-roller case. Fortunately, however, the two-roller system is simple enough that we can analyze the kinematic constraint and freedom requirements of the two rollers in a fairly straightforward manner without the aid of the web plane diagram.

Getting Uniform Web Tension

The web must be installed so that it is under uniform tension, irrespective of any conical shape in the web or rollers. (The web is said to have a conical shape if one edge is longer than the other.) Figure 8.13.1 is a schematic representation of the degrees of freedom needed to achieve this. The axis of roller A is fixed, but roller B has two freedoms. It must be able to translate along the X-axis and pivot about the Z-axis. The web provides constraint to the B roller in X and θ_z.

Figure 8.13.2 is an alternative and equally valid statement of the freedoms needed to establish uni-

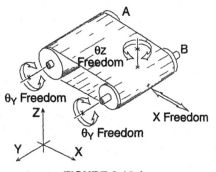

FIGURE 8.13.1

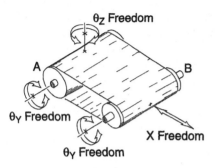

FIGURE 8.13.2

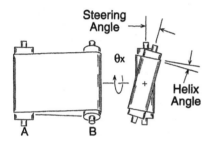

FIGURE 8.13.3

form web tension. Roller A would be free in θ_z, whereas roller B would be free in X. This may not turn out to be practical when reduced to hardware, but it certainly would work. For the purpose of this discussion, we will confine our attention to the configuration of Fig. 8.13.1.

Web "Cross-Track" Position

Next we consider how the web's lateral or "cross-track" position is determined. First, consider the effect of a small θ_x rotation of roller B with respect to roller A, as shown in Fig. 8.13.3. (Note that this is equivalent to and indistinguishable from a θ_x rotation of roller A with respect to roller B.) This results in the web "pivoting" (steering) on the surface of each roller so that it now wraps each roller at a small helix angle. As the rollers turn, the web advances with a small component of cross-track motion.

This is analogous to the apparent motion of the red stripe on a barber pole. Since the red stripe is wrapped helically around the barber pole, as the pole turns, the stripe advances axially. The apparent speed of the red stripe in the axial direction depends on its helix angle. Analogously, the web's cross-track velocity also depends on the helix angle of the web on the rollers. Because we have already seen that this helix angle depends on roller steering angle θ_x, we can conclude the following:

The web's cross-track *velocity* is controlled by the roller's steering angle θ_x.

Thus if we can sense or detect any error in the lateral position of the web edge, we can then use this information to control the rollers' steering angle, which, in turn, causes the web edge lateral position to move to (or stay in) its intended location.

Another approach to controlling the web's cross-track position is to simply constrain it with edge guides. Of course, we want to avoid overconstraining the web, so if we use an edge guide, the rollers must be rendered zero constraints.

Hardware Examples

The apparatus shown in Fig. 8.13.4 is designed to convey the web on two *axially compliant rollers*. A six-sided box structure serves as a rigid core. The left roller is mounted in fixed bearings. The right roller is

pivotally attached to spring-loaded side plates, permitting X and θ_z freedom for uniform web tension.
Pivoting paddle edge guides straddle the web as it
wraps one of the rollers.

Alternatively, one of the rollers can be castered in
order to achieve zero constraint. In the apparatus of
Fig. 8.13.5, the right roller is castered. It spins on a
curved shaft. The center of curvature of the curved
section of the shaft defines the roller's caster axis.

Pivoting paddles establish "hard" cross-track
boundaries for the web. Because the web is not permitted to move cross-track, any helix angle between
the web and roller forces the *roller* to move laterally.
Because the roller is castered, this produces an associated steering angle change, which, of course, must
be in a direction to *reduce* the helix angle. In other
words, the caster axis must be located upstream, not
downstream.

The force of the web edge against the paddles is
equal to the force required to cause the roller to move
laterally. This turns out to be practically nil. Because the
roller is already rotating on its shaft, only an infinitesimal lateral force is needed to get it to move axially.

In the apparatus shown in Fig. 8.13.6, the right
roller is not *free* to rotate about the caster axis, but is
instead driven about the caster axis by a servomotor.
Also, the pivoting paddle edge guides have been
replaced by a pair of optoelectronic edge sensors,
mounted slightly staggered in the Y direction. If the
web edge is between the two sensors (within the
"dead band"), no signal is sent to the servomotor. But
when the web edge goes outside the dead band, the
motor is turned on. The servomotor, through a lead
screw, causes the roller to rotate about its caster axis.
This has two immediate effects. First, the web edge is
carried back into the dead band, resulting in the motor
turning off again (negative feedback). Second, the
steering angle between the two rollers is changed in a
direction to reduce the web cross-track velocity that
caused the web edge to go outside the dead band in
the first place. After several such steering corrections,
the web cross-track velocity is reduced to nearly nil
and the servomotor only rarely comes on again.

Rotation about the caster axis in the above example produced two simultaneous effects:

1. Lateral (cross-track) shift between the sensors
   and the web edge, and
2. Steering axis correction.

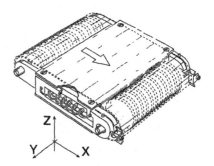

FIGURE 8.13.4

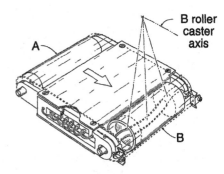

FIGURE 8.13.5

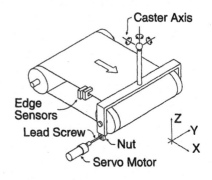

FIGURE 8.13.6

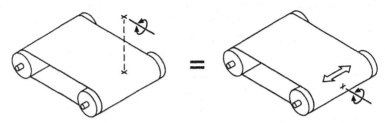

FIGURE 8.13.7

Indeed, we know that a rotation about the caster axis can be equivalently expressed as rotation about a parallel axis (the steering axis) coupled with an orthogonal translation (Y translation). (See section 3.9.)

This gives us the idea to redesign the apparatus of Fig. 8.13.6 to relocate the caster axis to the steering axis. (The steering axis is parallel to the caster axis but intersects the roller axis.) In relocating the caster axis to the steering axis, we also need to provide a *coupled* lateral motion of the edge sensors.

The apparatus of Fig. 8.13.8 accomplishes a result equivalent to that of the previous example (Fig. 8.13.6) by providing steering axis rotation and coupled cross-track translation between the web edge and sensors. The sensors are mounted so that they will move in the cross-track direction whenever the roller steering angle is changed.

The magnitude of the coupled cross-track translation is $T = l \times R$, where l is the caster radius being mimicked and T is the caster rotation.

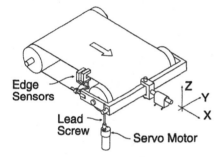

FIGURE 8.13.8

8.14 ■ CROWNED ROLLERS

In the previous section, we investigated the behavior of a two-dimensionally rigid web being conveyed on a pair of rollers. We discovered that by steering the two rollers about their mutual steering axis, we can change the helix angle of the web's wrap around the rollers. We found that the web's rate of cross-track motion was proportional to this helix angle. The analogy to the apparent axial motion of the red and white stripes on a rotating barber pole was drawn. Let us now expand on this analogy a little.

Consider the hypothetical machine shown in Fig. 8.14.1, which is used to manufacture barber poles. This machine applies a red stripe (a web) helically to a rotating white pole (a roller). The stripe is

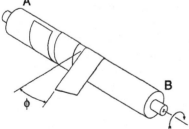

FIGURE 8.14.1

applied at an angle φ, where φ is the angle between the edge of the stripe (in the span approaching the pole) and a line perpendicular to the axis of the pole. φ Is also the helix angle of the stripe on the pole. The stripe is initially attached at the end of the pole labeled "A" and is then wound onto the pole until it reaches the end labeled "B." Clearly, by controlling the angle of the web span as it approaches the pole, we control the helix angle φ, and thus control cross-track velocity of the stripe. We can use this model as we try to understand the behavior of any web approaching a roller at an angle φ. It should be clear, by analogy, that when a web approaches a roller at an angle φ, the web will have a component of cross-track velocity toward the same side as that from which it approached the roller.

Now we will investigate the use of a crowned roller, such as the one shown in Fig. 8.14.2, for the purpose of tracking a web on a two-roller apparatus. In this case, the crowned roller has the shape of two cones (frustums) positioned base-to-base. As was mentioned in section 8.1, crowned rollers are useful in the conveyance of two-dimensionally compliant webs.

To understand the mechanism of crowned rollers, refer to Fig. 8.14.3, which shows a section of a web positioned entirely on one half of the crowned roller. Notice that because of the roller's conical shape, the web is forced into a condition where it approaches the roller at some angle φ. From our previous discussion about the barber pole, it should be clear that this approach angle will result in the web having a component of cross-track velocity that will cause it to move toward the large end of the cone. For the web to pass around both rollers, it must follow a curved path from the cylindrical roller to the conical (crowned) roller, as in Fig. 8.14.4. Obviously, it must be two-dimensionally compliant to follow this path. This curved path of the web is essential to maintaining the nonzero approach angle of the web to the crowned roller. If, for example, a stiff (two-dimensionally rigid) web is used instead, the web would be incapable of following the curved path and would therefore be unable to achieve the nonzero approach angle to the conical section of the crowned roller. The crown would have no effect on a stiff web. But with a compliant web, the approach angle is maintained and the web continues to move in the cross-track direction until it reaches the largest diameter of the crowned roller.

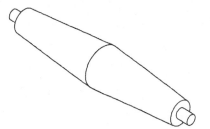

FIGURE 8.14.2

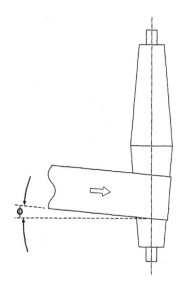

FIGURE 8.14.3

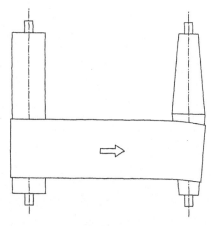

FIGURE 8.14.4

CHAPTER SUMMARY

In this chapter Exact Constraint Design principles were applied to the design of web conveyance machinery. Using familiar analogies, we learned the exact nature of the connection between each conveyance roller and the web. Then, with the aid of a web plane diagram we got a composite view of the total constraint pattern applied to the web. We then saw clearly why traditionally designed web conveyance hardware must rely so heavily on accurate alignments in order to work. More importantly, we acquired a tool that enables us to design web conveyance hardware that does *not* overconstrain the web and which can be expected to work reliably irrespective of part tolerances and web variations.

Index

CPSIA information can be obtained at www.ICGtesting.com
Printed in the USA
BVOW06*0020270816

460332BV00008B/139/P